DK 动物成长奥秘
看！我在长大（中英双语版）

海龟

英国 DK 公司 ◎ 编
褚诗语 刘润琦 闫书旻 张晴焱 ◎ 译
鹰之舞 沈成 ◎ 审

人民邮电出版社
北京

Original Title: Turtle
Copyright © Dorling Kindersley Limited, 2006
A Penguin Random House Company

本书简体中文版授权由人民邮电出版社独家出版，仅限于中国境内（不包括香港、澳门、台湾地区）销售。未经出版者书面许可，不得以任何方式复制或发行本书中的任何部分。

For the curious
www.dk.com

目录 Contents

4~5
我是一只绿海龟

6~7
我的爸爸和妈妈

8~9
妈妈藏起蛋宝宝

10~11
是时候破壳而出啦

12~13
大海，我来啦

14~15
在海藻里躲猫猫

16~17
洗澡时间到

18~19
我长大啦

20~21
生命循环，周而复始

22~23
我世界各地的朋友

24
词汇表

我是一只绿海龟
（I'm a green sea turtle）

我是一只绿海龟宝宝。我一生的大部分时间都生活在海洋（ocean）。我在大海里畅游，和其他的小伙伴（friend）一起做游戏。我只能用肺呼吸，所以需要经常浮到海面呼吸新鲜的空气。我喜欢吃各种各样的海草，有时也会捕食一些小动物（small animal）来换换口味。

海龟像全副武装的骑士一样，它们身披厚厚的"铠甲"，"铠甲"可以保护它们的前胸（chest）和后背（back）。

鳍状四肢上覆盖着坚硬的鳞片（scale）。

海龟的眼睛（eye）也被厚厚的眼皮（eyelid）保护着。

像人类一样，海龟用两个鼻孔（nostril）呼吸！

我的分布范围非常广，世界各地的温暖海域中都可以找到我的身影。

海龟没有牙，但它的嘴像鸟嘴（beak）一样，又坚硬又锋利，可以咬碎食物。

海龟的皮肤（skin）富有弹性，这样它的头和四肢就可以伸缩自如。

我的爸爸和妈妈
（My dad and mum）

我的爸爸和妈妈在海里遨游时一见钟情。交配后，妈妈会产卵（egg）4~5窝。妈妈一般半个月产一窝蛋。

这是我的爸爸。

海龟爸爸用四肢（flipper）紧紧地抱住海龟妈妈。

这是我的妈妈。

交配后，海龟夫妻就各奔东西了。

游向岸边

海龟会游到距离栖息地上百甚至上千千米的地方交配、产卵。

嘿呦，加油！

海龟妈妈依靠自己强壮有力的（strong）前肢从海里爬到沙滩上。拖着沉重的身体在沙滩上爬行是件辛苦的事情，海龟妈妈通常选择在凉爽的清晨或者夜晚（morning or night time）行动。

妈妈藏起蛋宝宝
（Mum buries her eggs）

一番精挑细选后，海龟妈妈选好了地点来打造"婴儿房"。推开沙子，挖出一个深深的洞，然后就可以安心产卵啦！沙子又软又暖（warm），蛋壳（shell）也是软软的，一个接一个的蛋掉在"沙床"上，海龟妈妈也不必担心蛋壳被摔破。

安然无恙

产卵后，海龟妈妈用后肢推动沙子盖住蛋宝宝，把它们小心翼翼地藏起来。海龟妈妈每年（every year）都会选择在同一片沙滩（beach）上产卵。

建造这样一个"婴儿房",再加上产卵,都是既费时又费力的事儿,"单枪匹马"的海龟妈妈通常要耗费好几小时才能完成这项大工程。

挖呀挖,海龟妈妈用自己强而有力的四肢快速地推开沙子。

"婴儿房"的空间很大(big),能装下60~120个蛋宝宝呢!

是时候破壳而出啦
（It's time to hatch out）

在"婴儿房"里呼呼大睡两个月（two months）后，我们要破壳而出啦！海龟宝宝陆续啄破蛋壳，一起努力掀开盖在身上的厚厚的"沙被"，争先恐后地爬上沙滩。

通往外界的路上阻碍重重，它们至少需要整整一周（a whole week）才能完成这一艰巨的任务。

撸起袖子加油干！

嘘！这只螃蟹（crab）似乎察觉到了沙滩下的动静，早已饥肠辘辘的它，迫不及待地要捕食啦！小海龟加油（hurry up）啊！可千万别被它抓住！

海龟小知识

- 🐢 小海龟有一颗小而硬的牙齿，称作"卵齿"。小海龟依靠"卵齿"破壳而出。

- 🐢 刚出生的海龟宝宝大约只有5厘米。

- 🐢 如果"婴儿房"里的温度比较高，那么孵出来的都是雌性小海龟；温度低的话，则出生的都是雄性小海龟。

大海,我来啦
(I'm off to the sea)

历经千辛万苦,我和兄弟姐妹(brothers and sisters)终于爬到了沙滩上。我们争先恐后地冲向大海,各奔东西。

海里更安全。快快快!越快越好!让我们一起奔向大海。

沙滩上的鸟（bird）、蜥蜴（lizard）和其他动物都虎视眈眈地等着吃大餐呢！

海龟宝宝也很怕太阳暴晒，于是它们等到凉快（cool）的时候，就飞快地逃离危机四伏的沙滩，冲向大海。

为安全（safety）起见，聪明的海龟宝宝往往会集体行动，抓准时机一起离开，爬向大海。

在海藻里躲猫猫
（Seaweed hide and seek）

现在我已经逃离沙滩来到大海了。但大海里依然危机四伏，我只能选择躲藏在海藻（seaweed）里。这里不仅安全，还有丰富的食物。

在之后的岁月里，小海龟随波逐流，海藻就是它们的藏身之所。

这只小海龟已经两个月大了。为了能随时浮出海面来自由地呼吸（breathe），它会待在离海面很近的地方。

海龟的菜单

🐢 成年海龟主要吃一些生长在大海深处的海草，但是海龟宝宝主要以在浅海处活动的浮游生物为食。水母和海绵也是海龟宝宝喜爱的美食。

浮游生物

洗澡时间到
（It's time to get clean）

绿海藻

我的个头越长越大，胆子也越来越大，出门在外的时间越来越长，活动范围也越来越广。渐渐地，我的壳上盖满了藻类（algae）。这些藻类对我的身体可没有好处，所以我一定得把它们清理掉。瞧，我和朋友们正排着队等待小鱼（fish）来帮我们清洁身体呢！

在"清洁站"里,鱼儿们吃饱了肚子,海龟也变得干干净净了。

无"藻"一身轻……伙计们,再见!

快速清洁只需要几分钟就可以完成,深度清洁则要耗费几小时。

清洁小知识

- 黏附在海龟壳上的海藻不仅会减缓海龟在水中的游泳速度,而且更可怕的是会导致它们生病。

- 在清洁站里,海龟放松地将四肢完全舒展,这样清洁工小鱼儿就可以把它们身体的每一个部位清洗得干干净净。

- 有时候,助理小虾也会来帮忙清洗哦!

我长大啦
（I'm a big turtle now）

我15岁啦！现在的我终于可以在大海里畅游了！但大多数时候，我还是更喜欢待在靠近陆地（land）的地方，这里有丰富的食物（food）供我享用。

海龟们是名副其实的游泳健将（swimmer）。

海龟们用肺呼吸，所以需要时不时地浮出海面换气。

乐享水下生活

成年海龟是潜水能手，每换一次气，最多可以在海里待上5小时。

绿色套餐

成年海龟是素食主义者，它们只吃绿色植物，比如海草（sea grass）和海藻等。海龟一生中的大部分时间都是在浅海（shallow areas of the sea）中觅食。

嘿，这只海龟正琢磨着找个地方美美地睡上一觉，那边的礁石洞穴（rock cave）正合适。

生命循环，周而复始
The circle of life goes round and round

现在你知道我怎样成长为一只成年海龟了吧！

Now you know how I turned into a grown-up sea turtle.

大海任我游！伙伴们，我要出发啦！再见！

我世界各地的朋友
My friends from around the world

尽管我和我的朋友们生活在不同的地方，有的在淡水里，有的在海洋里，但"龟"是我们共同的名字。

长颈龟（Snake-Necked Turtle）遍布澳大利亚的淡水河。

玳瑁（Hawksbill Turtle）的踪影在全球的温暖海域里很常见。

我是一只棱皮龟。

善于捕鱼的大鳄龟（Alligator Snapping Turtle）以美国南部河流为家。

生活在大海里的丽龟，因为它具有独特的绿色龟壳，又被称为"橄榄龟"（Olive Ridley Turtle）。

我是一只小蠵龟。

生活在大洋里的棱皮龟（Leatherback Turtle），体形巨大，甚至可以长到一辆小轿车那么大。

海龟小知识

- 吃了绿色的海藻后，绿海龟身上的脂肪也变成了绿色，它也因绿色的脂肪而得名。

- 海龟的寿命可以超过80年。

- 棱皮龟与众不同，虽然没有硬硬的铠甲，但它那厚厚的皮肤有着骨骼的支撑，犹如一面皮质的盾牌，这样的身体构造可以起到保护作用。

词汇表 Glossary

鳍状肢
Flipper
海龟的前肢，形状扁平，适于划水。

孵化
Hatch
海龟宝宝破壳而出的过程。

喙
Beak
海龟嘴上部的坚硬部分，用于进食。

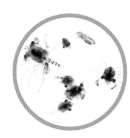

浮游生物
Plankton
微小的海洋动物，是海龟的食物。

鳞片
Scales
坚硬、扁平的鳞甲，构成海龟的壳和皮肤。

礁石
Reef
由珊瑚、岩石和沙子组成的质地坚硬的海中山脊。

致谢 Acknowledgements

感谢以下人员及机构提供图片：
(Key: a=above; c=centre; b=below; l=left; r=right; t=top)

1 SeaPics: James D.Watt. 2-3: SeaPics: Doug Perrine. 4-5 Alamy: Carlos Villoch. 4 Oxford Scientific Films: Gerard Soury tl; Alamy: Carlos Villoch bl. 5 Oxford Scientific Films cr. 6 Nature Photo Library: Doug Perrine. 7: Getty Images: David Fleetham t; Jim Angy b. 8: Science Photo Library: Alexis Rosenfeld l. 9 Getty Images: Cousteau Society. 10-11 James L. Amos. 10 Alamy: Aqua Image c; Image Quest Marine: Tim Hellier bl; Still Pictures: Kevin Aitkin ctr. 11 Corbis: Kevin Schafer cl; Getty Images: Tim Laman tr; Natural Visions: Soames Summerhays br. 12 Frank Lane Picture Agency: Frans Lanting/Minden Pictures bl; SeaPics: Doug Perrine tl; NHPA: B. Jones & M.Shimlock. 14-15 Getty Images: Bill Curtsinger. 15 Science Photo Library: Alexis Rosenfeld tr. 16 Getty Images: A Witte/C Mahaney b; Oxford Scientific Films: Photolibrary tl. 17 Maui Sea Life: Doug & Kerry Pilot. 18-19 Getty Images: Jeff Hunter. 18 Getty Images: Michael Gilbert tl. 19 Oxford Scientific Films: Photolibrary br; SeaPics: Doug Perrine t. 20 Alamy: M. Timothy O'Keefe bl; Alamy: Aqua Image crb; Corbis: Kevin Schafer cra; Image Quest Marine tc & tcr; Image Quest Marine: James D. Watt clb; Natural Visions: Soames Summerhays crbb; NHPA: Linda Pitkin cl; Oxford Scientific Films/Photolibrary: Gerard Soury bc; Science Photo Library: Alexis Rosenfeld c; Lumigenic: Mark Shargel tl. 21 Oxford Scientific Films/Photolibrary. 22-23 Still Pictures: Kelvin Aitken cb. 22 Alamy: Michael Patrick O'Neill tr. 23 Alamy: CuboImages srl; Alfio Giannotti rc; NHPA: Martin Wendler tr. 24 Alamy: Carlos Villoch bl; Corbis: Kevin Schafer tr; Getty Images: Jeff Hunter br; Oxford Scientific Films/Photolibrary cl; SeaPics: James D. Watt tl.

其他图片版权属于多林·金德斯利公司。欲了解更多信息请访问DK Images网站。

Glossary

Flipper
The turtle's arm. It is flat and shaped for swimming.

Hatch
When the baby sea turtle pecks its way out of its egg.

Beak
The hard upper part of the turtle's mouth, used for eating.

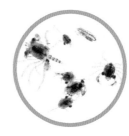
Plankton
Tiny sea animals that are food for the sea turtle.

Scales
Hard, flat plates that make up the turtle's shell and skin.

Reef
A hard ridge made up of coral, rocks, and sand.

Acknowledgements
感谢以下人员及机构提供图片：
(Key: a=above; c=centre; b=below; l=left; r=right; t=top)

1 SeaPics: James D.Watt. 2-3: SeaPics: Doug Perrine. 4-5 Alamy: Carlos Villoch. 4 Oxford Scientific Films: Gerard Soury tl; Alamy: Carlos Villoch bl. 5 Oxford Scientific Films cr. 6 Nature Photo Library: Doug Perrine. 7: Getty Images: David Fleetham t; Jim Angy b. 8: Science Photo Library: Alexis Rosenfeld l. 9 Getty Images: Cousteau Society. 10-11 James L. Amos. 10 Alamy: Aqua Image c; Image Quest Marine: Tim Hellier bl; Still Pictures: Kevin Aitkin ctr. 11 Corbis: Kevin Schafer cl; Getty Images: Tim Laman tr; Natural Visions: Soames Summerhays br. 12 Frank Lane Picture Agency: Frans Lanting/Minden Pictures bl; SeaPics: Doug Perrine tl; NHPA: B. Jones & M. Shimlock. 14-15 Getty Images: Bill Curtsinger. 15 Science Photo Library: Alexis Rosenfeld tr. 16 Getty Images: A Witte/C Mahaney b; Oxford Scientific Films: Photolibrary tl. 17 Maui Sea Life: Doug & Kerry Pilot. 18-19 Getty Images: Jeff Hunter. 18 Getty Images: Michael Gilbert tl. 19 Oxford Scientific Films: Photolibrary br; SeaPics: Doug Perrine t. 20 Alamy: M. Timothy O'Keefe bl; Alamy: Aqua Image crb; Corbis: Kevin Schafer cra; Image Quest Marine tc & tcr; Image Quest Marine: James D. Watt clb; Natural Visions: Soames Summerhays crbb; NHPA: Linda Pitkin cl; Oxford Scientific Films/Photolibrary: Gerard Soury bc; Science Photo Library: Alexis Rosenfeld c; Lumigenic: Mark Shargel tl. 21 Oxford Scientific Films/Photolibrary. 22-23 Still Pictures: Kelvin Aitken cb. 22 Alamy: Michael Patrick O'Neill bl. 23 Alamy: CuboImages srl; Alfio Giannotti rc; NHPA: Martin Wendler tr. 24 Alamy: Carlos Villoch bl; Corbis: Kevin Schafer tr; Getty Images: Jeff Hunter br; Oxford Scientific Films/Photolibrary cl; SeaPics: James D. Watt tl.

其他图片版权属于多林·金德斯利公司。欲了解更多信息请访问DK Images网站。

Alligator Snapping Turtles eat fish and live in rivers in the southern part of the United States of America.

The Olive Ridley Turtle lives in the sea and gets its name from its green shell.

I'm a young Logger Head Turtle.

Leatherback Turtles live in the ocean and can grow as big as a small car.

Turtle facts

- The Green Sea Turtle gets its name from the colour of its body fat, which is green from the algae it eats.

- Sea turtles can live to be more than 80 years old.

- Instead of a hard shell, the Leatherback Turtle has a thick skin that is supported by bones.

My friends from around the world

Some of my friends live in freshwater rivers and lakes and some live in the salty ocean with me. But we are all turtles.

Snake-Necked Turtles live in freshwater rivers all over Australia.

Hawksbill Turtles live in warm seas all around the world.

I'm a Leatherback Turtle.

Bye bye, I'm off to swim in the seas.

The circle of life goes round and round

Now you know how I turned into a grown-up sea turtle.

Eat your greens

Adult sea turtles eat only plants. They eat seaweed, sea grass, and algae. They spend most of their time looking for food in shallow areas of the sea.

This turtle is looking for a rock cave to sleep in.

I'm a big turtle now

I'm 15 years old. I'm finally big enough to swim anywhere in the ocean on my own. Most of the time I like to stay close to land, where there is a lot of food to eat.

Sea turtles need to surafce for air.

Sea turtles are very strong swimmers.

Underwater life
Adult sea turtles can stay under water for up to five hours before taking a breath of air.

At the cleaning station, the fish get food, and the turtles get cleaned.

All clean... See you later guys.

Sometimes cleaning only takes a few minutes — other times it can take hours.

Cleaning facts

🐢 The algae on the turtles' shells slows them down in the water and can cause illness.

🐢 At the cleaning station, the turtles stretch out so the fish can reach every spot.

🐢 Sometimes shrimp also help with the cleaning.

17

It's time to get clean

Green algae

As I grow bigger and swim around more, my shell gets covered in algae. Algae is not good for me, so I need to keep clean. My friends and I line up and wait for fish to eat up all the algae on our shells.

Turtle treats

Adult sea turtles eat mainly sea-grasses that grow deeper down, but baby sea turtles eat all sorts of tiny animals that live near the surface. These are called plankton. Jellyfish and sponges also make a tasty treat for the young turtles.

Plankton

This turtle is two months old. She stays close to the surface so she can breathe.

Seaweed hide and seek

Now that I am in the ocean, I stay safe by hiding in the seaweed. This will be my home until I am bigger. I eat small animals that live in the seaweed.

The turtles will spend many years floating on seaweed.

On land, birds, lizards, and other animals will try to catch the baby turtles.

The hatchlings wait until it's cool to run for the sea.

The hatchlings all leave their nest at the same time, for safety.

I'm off to the sea

My brothers and sisters and I work together to dig our way to the surface. Then we all rush for the sea. Once we are in the water, we all swim away.

Hurry up! It's safer in the sea, let's get there as quickly as we can.

This crab can feel something moving under the sand. Hurry up turtles — he's starting to dig for his dinner!

Turtle facts

- The turtles use a tiny, hard tooth, called an egg tooth, to help them break open their shell.
- The baby sea turtles are about 5 cm (2 in) long.
- If the temperature in the nest is hot, all the hatchlings will be girls. If it is cooler, they will all be boys!

It's time to hatch out

After two months under the sand, we hatch out of our shells. Once everyone has hatched, we all work together to dig to the surface.

Digging through the sand is very hard work. It can take the turtles a whole week to dig to the surface.

Let's get digging.

Digging and laying eggs is hard work. It can take the turtle a few hours to dig the nest and lay the eggs.

Mum's powerful flippers flick and fling the sand away.

The nest will be big enough to hold between 60 and 120 eggs.

9

Mum buries her eggs

My mum crawls up on to the sand to lay her eggs. She digs a hole and lays them one at a time. Our eggs have soft shells so they do not break when they fall. The sandy nest will protect us and keep us warm.

Safe and sound
After laying the eggs, the turtle covers them with sand using her back flippers. Turtles lay their eggs on the same beach every year.

Swimming to shore

Sea turtles can travel hundreds or thousands of miles from the place where they live to the place where they mate and lay eggs.

Heave ho, up we go...

The female turtle uses her strong front flippers to drag herself out of the water and on to the sand. It's hard work and she usually waits until morning or night time, when it's cool.

My dad and mum

My dad and mum met while swimming in the sea. After mating, mum will lay four or five nests full of eggs. She will lay one nest every two weeks.

The male turtle uses his fippers to hold on to the female.

This is my dad.

This is my mum.

After mating, the turtles do not stay together.

The turtle's eyes are protected by thick eyelids.

Turtles breathe air through two nostrils, just like us!

I live in oceans all over the world.

Sea turtles tear their food with a sharp beak.

Soft, bendy skin allows the turtle's head and flippers to move.

I'm a green sea turtle

I'm a green sea turtle. I swim in the ocean, but I come to the surface to breathe air. I eat plants and small animals from the sea and spend most of my life under the water with my friends.

A hard shell covers the turtle's back and chest.

The flippers are covered in tough scales.

18~19
I'm a big turtle now

20~21
The circle of life

22~23
My friends from around the world

24
Glossary

DK WATCH ME GROW
TURTLE

Contents

4~5
I'm a green sea turtle

6~7
My dad and mum

8~9
Mum buries her eggs

10~11
It's time to hatch out

12~13
I'm off to the sea

14~15
Seaweed hide and seek

16~17
It's time to get clean

DK WATCH ME GROW
APES

Contents

4~5
I'm an orang-utan

6~7
Learning to climb

8~9
I'm a gorilla

I'm an orang-utan

I was born in a green, leafy jungle. My mum and I live up in the trees. I hang on tight to my mother while she swings from tree to tree.

Hold on tight.

The baby will stay with its mother for about eight years.

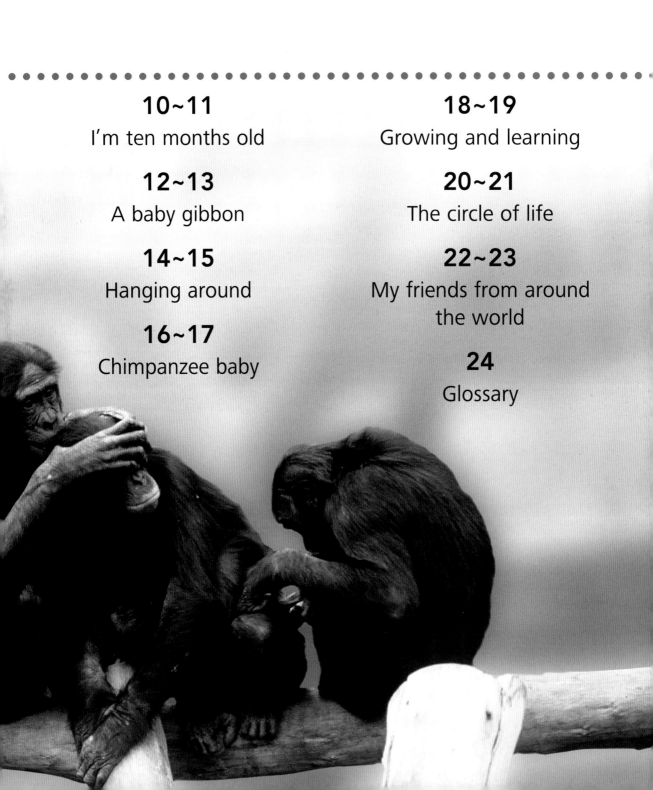

10~11
I'm ten months old

12~13
A baby gibbon

14~15
Hanging around

16~17
Chimpanzee baby

18~19
Growing and learning

20~21
The circle of life

22~23
My friends from around the world

24
Glossary

Orang-utans learn to climb when they are around three years old.

We're just hanging around.

Fruit for breakfast

This orang-utan is eating a jungle fruit called a durian. Orangutans love to eat fruit, but they also eat leaves, bark, flowers, and insects.

Learning to climb

I'm three years old. My brothers and sisters are teaching me how to climb trees on my own, but I am careful to stay close to mum.

Orang-utans have very strong hands and feet.

Hey! look at me, I can climb!

Orang-utan's teeth are designed for eating fruit.

Orang-utan facts

- One adult orang-utan is about as strong as eight adult humans.
- The arms of an adult orang-utan can be more than 2.5 m (7 ft) long when stretched out.
- Orang-utans use many tools. Some orang-utans use leaves as napkins, to wipe food off their faces.

Be careful, little brother!

Family life
Orang-utans live with their mums until they are about 8 years old, when they go off to live on their own.

I'm a gorilla

I live with my family in a big forest. I'm only four months old and I can't walk yet so I hang on to my mum. I cling to mum's fur while she looks for some tasty food to eat.

A free ride
This newborn gorilla will cling to its mother's fur for around five months. After that, the baby will ride on its mother's back or shoulders.

Are we there yet?

A gorilla's day
Gorillas spend their mornings and evenings looking for food and eating. They spend the middle part of the day sleeping, playing, or grooming.

I'm ten months old

I can walk on my own now, but whenever I get tired I hitch a ride with my older brother. I spend most of the day with my brothers and sisters.

Gorillas like to take a nap in the middle of the day.

Sweet dreams

When they go to sleep, gorillas make a nest out of leaves and branches. When they wake up, they eat the leaves and have breakfast in bed.

I also rest with my brothers and sisters.

A baby gibbon

I was born high up in the trees in a tropical rainforest. I live with my mother, my father, and my brothers and sisters. My family spend most of their time high up in the trees.

I can't wait until I'm old enough to swing through the trees.

Gibbons learn to swing through the trees when they are about one year old.

Look how far I can stretch!

Gibbon facts

- Gibbons sleep sitting up.
- Gibbons are very good at walking on two feet, just like humans.
- Gibbons eat fruit, leaves, and insects, but their favourite food is ripe figs.

Hanging around

My family and I move through the trees by swinging from branch to branch. My long arms and hands help me to hang on to the branches as I move.

Gibbons use their curved hands, feet, and fingers to hook on to branches as they swing from tree to tree.

My brother can swing all the way through the

Hanging on with just one hand is easy for a gibbon!

A gibbon can leap about 10 m (30 ft) from tree to tree.

forest.

Call of the wild

Each gibbon family makes up its own songs and sings them every day. The songs are very loud and tell other gibbons to stay away.

Chimpanzee baby

I was born in a forest. I live here with my family and all of our friends. There are always plenty of other chimps for me to play with.

New arrival
The newborn chimp is carried by her mother until she is strong enough to cling to her mother's fur.

Climbing is hard, hanging on is easy.

Chimps eat fruit, insects, honey, flowers, leaves, nuts, and small animals.

Keeping clean

Chimpanzees spend a lot of their time grooming each other. Grooming helps keep the chimps' fur clean and is also a way to make friends.

Baby chimps spend almost all of their time with their mother.

All this fun and play has made me sleepy!

Growing and learning

There are so many things to learn before I am all grown up! My sisters and brothers teach me what kinds of food are good to eat. They also show me how to use tools, such as sticks, to reach the food.

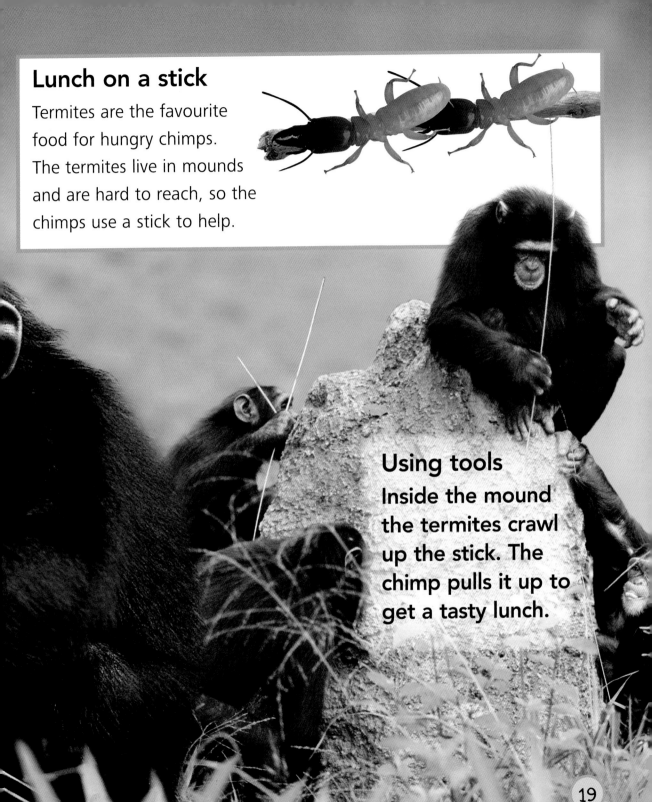

Lunch on a stick

Termites are the favourite food for hungry chimps. The termites live in mounds and are hard to reach, so the chimps use a stick to help.

Using tools

Inside the mound the termites crawl up the stick. The chimp pulls it up to get a tasty lunch.

The circle of life goes round and round

...baby to a gorilla

Now you know how we grew from a ...

...baby to a gibbon

... baby to an orangutan

Bye bye, I'll miss you.

... baby to a chimpanzee

My friends from around the world

The Borneo Gibbon lives only on the tropical island of Borneo.

Siamang Apes live in tropical forests in southeast Asia.

The Mountain Gorilla lives high up in the mountains in central Africa.

I'm tired,

The Bonobo lives in Africa and is also called the pygmy chimpanzee.

My ape friends live in jungles and forests all around the world.

The Buff-Cheeked Crested Gibbon has yellow or tan fur.

The Agile Gibbon lives in the tops of jungle trees.

it's time to sleep.

Ape facts

 Gibbons swing through the trees at about 56 km/h.

 Chimpanzees don't like to be in water and can't swim.

 Gorillas live together in groups of about 20, called troops.

 The word "orang-utan" means forest person.

Glossary

Nest
A place some apes make for sleeping, made out of leaves.

Fur
The soft hair that keeps the ape warm and protects it.

Grooming
When apes use their hands to clean and tidy each other's fur.

Tool
An object used for a special job, like a stick or a rock.

Rainforest
A place with many big trees where it rains a lot.

Troop
A group of animals living together. Chimps live in troops.

Acknowledgements

感谢以下人员及机构提供图片：

(Key: a=above; c=centre; b=below; l=left; r=right; t=top)
1 Alamy: Juniors Bildarchiv. 2-3 Ardea: Kenneth W. Fink. 2 Getty Images: David Allan Brandt c; Catherine Ledner cr. FLPA: Frans Lanting cl. 3 Alamy: Steve Bloom Images. 4 Getty Images: Heinrich van der Berg. 5 Getty Images: Steven Raymer b; Art Wolfe t. 8-9 Getty Images: Tom Brakefield. 8 Ardea: John Cancalosi l. 10 Alamy: Martin Harvey 11 Alamy: Martin Harvey tr, c; NHPA cb. 12 FLPA: Jurgen & Christine Sohns. 13 Ardea London: M. Watson. 14 FLPA: Terry Whittaker tr; NHPA Gerard Lacz l. 15 Corbis: Martin Harvey tl; FLPA: Terry Whittaker r; Getty Images: Manoj Shah bl; NHPA: Martin Harvey tl. 16-17 Corbis: Steven Bein c. 16 Photolinrary: Richard Packwood bl;

17 Photolibrary: Richard Packwood bl; Steve Bloom Images cl. 17 Corbis: Mary Ann McDonald br; Gallo Images tc. 18-19 Getty Images: Digital Vision. 19 Steve Bloom Images r. 20-21 FLPA: Frans Lanting. 20 Alamy: Images of Africa Photobank cr. Ardea London: John Cancalosi c; M. Watson cbr. Corbis: William Manning c. 21 Corbis: Stan Osolinski cr; Lynne Renee cl. Getty Images: Paula Bronstein bl; Gallo Images tl. IPN stock: Catherine Ledner cla. Photolibrary: Stan Osolinski clb. Zefa: T. Allofs cl. 22-23 Alamy: Martin Harvey b. 22 Alamy: Chris Fredriksson tr, David Moore cl. Steve Bloom Images: bl. IPN stock: Catherine Ledner tl. 23 Alamy: Jack Cox-Travel Pics Pro tc. Photolibrary: cr. Steve Bloom Images: tr

其他图片版权属于多林·金德斯利公司。欲了解更多信息请访问DK Images网站。

词汇表 Glossary

巢
Nest
用树叶搭建的、用于睡觉的地方。

皮毛
Fur
用来保暖并起保护作用的皮肤和毛发。

梳毛
Grooming
互相清洁、整理毛发。

工具
Tool
为了完成某一特定任务而使用的物体，比如树枝或石头。

雨林
Rainforest
雨量充沛、树木繁茂的地方。

族群
Troop
同种动物生活在一起。如黑猩猩就是生活在族群里的。

致谢 Acknowledgements

感谢以下人员及机构提供图片：

(Key: a=above; c=centre; b=below; l=left; r=right; t=top)
1 Alamy: Juniors Bildarchiv. 2-3 Ardea: Kenneth W. Fink. 2 Getty Images: David Allan Brandt c; Catherine Ledner cr. FLPA: Frans Lanting cl. 3 Alamy: Steve Bloom Images. 4 Getty Images: Heinrich van der Berg. 5 Getty Images: Steven Raymer b; Art Wolfe t. 8-9 Getty Images: Tom Brakefield. 8 Ardea: John Cancalosi l. 10 Alamy: Martin Harvey 11 Alamy: Martin Harvey tr, c; NHPA cb. 12 FLPA: Jurgen & Christine Sohns. 13 Ardea London: M. Watson. 14 FLPA: Terry Whittaker tr; NHPA Gerard Lacz l. 15 Corbis: Martin Harvey tl; FLPA: Terry Whittaker r; Getty Images: Manoj Shah bl; NHPA: Martin Harvey tl. 16-17 Corbis: Steven Bein c. 16 Photolinrary: Richard Packwood bl; 17 Photolibrary: Richard Packwood bl; Steve Bloom Images cl. 17 Corbis: Mary Ann McDonald br; Gallo Images tc. 18-19 Getty Images: Digital Vision. 19 Steve Bloom Images r. 20-21 FLPA: Frans Lanting. 20 Alamy: Images of Africa Photobank cr. Ardea London: John Cancalosi c; M. Watson cbr. Corbis: William Manning c. 21 Corbis: Stan Osolinski cr; Lynne Renee cl. Getty Images: Paula Bronstein bl; Gallo Images tl. IPN stock: Catherine Ledner cla. Photolibrary: Stan Osolinski clb. Zefa: T. Allofs cl. 22-23 Alamy: Martin Harvey b. 22 Alamy: Chris Fredriksson tr, David Moore cl. Steve Bloom Images: bl. IPN stock: Catherine Ledner tl. 23 Alamy: Jack Cox-Travel Pics Pro tc. Photolibrary: cr. Steve Bloom Images: tr

其他图片版权属于多林·金德斯利公司。欲了解更多信息请访问DK Images网站。

我的朋友生活在世界各地的雨林和森林里。

红颊长臂猿（Buff-Cheeked Crested Gibbon）的体毛通常呈黄色或棕褐色。

白须长臂猿（Agile Gibbon）生活在丛林的树冠上。

该睡觉了！

猿类小知识

 长臂猿在树林里荡跃的速度可达56千米/小时。

 黑猩猩不会游泳，也不喜欢待在水里。

 大猩猩成群生活，每个群体约有20只。

 红毛猩猩在马来语和印尼语里叫作"orang-utan"，意思是"森林中的人"。

我世界各地的朋友
My friends from around the world

灰长臂猿（Borneo Gibbon），别称婆罗洲长臂猿，仅生活在婆罗洲的热带岛屿上。

合趾猴（Siamang Apes），也称大长臂猿，生活在东南亚的热带雨林里。

山地大猩猩（Mountain Gorilla）生活在中非的高山上。

我好累啊！

倭黑猩猩（Bonobo），又名侏儒黑猩猩，生活在非洲。

……宝宝长成红毛猩猩

宝宝长成大猩猩……

再见，我会想你的！

生命循环，周而复始
The circle of life goes round and round

……宝宝长成大猩猩

现在你知道我们
怎样从……

长斗长成大猩猩……

白蚁串

对饥肠辘辘的黑猩猩来说,白蚁(termite)可是无上的美味!不过,白蚁生活在大土堆一样的蚁巢里,想吃到它们是有难度的。这可难不倒聪明的黑猩猩,它用树枝或木棍智取猎物,将白蚁一举擒获!

使用工具

黑猩猩把树枝或木棍伸进白蚁穴内,白蚁纷纷爬上去,黑猩猩将树枝或木棍轻轻抽出,美味的白蚁串就到手啦!

成长和学习
（Growing and learning）

在长大成年之前,我还有好多本领需要学习啊!哥哥姐姐正在教我如何辨别食物,还有如何借助草秆或细树枝这样的工具（tool）获取食物。

保持干净

黑猩猩会花很多时间和同伴互相梳理毛发,这样不但能让身体保持干净(clean),而且有助于它们结交朋友(make friends)、加深感情。

每时每刻,黑猩猩幼崽都和妈妈寸步不离。

真好玩!不过我有点困了。

黑猩猩幼崽
（Chimpanzee baby）

我和家人、朋友（friend）生活在广袤的森林，我们常常聚集在一块儿，从不缺少玩伴。

生命之初

刚出生的黑猩猩幼崽没有行动能力，一直由妈妈抱着，直到它足够强壮，能抓住妈妈身上的皮毛。

爬树可真累啊，抱着树，偷会儿懒吧！

黑猩猩的食性较杂，食物包括水果（fruit）、昆虫、蜂蜜、花朵、树叶、坚果（nut），以及某些小动物。

长臂猿全力一跃,能跃到10米外的另一棵树上。

丛林之歌

每个长臂猿家庭都有自己独创的歌曲(song),并每天咏唱(sing),歌声嘹亮,警告其他长臂猿:"不要靠近!"

林间荡跃
（Hanging around）

我们一家每天都在树林里自由穿梭，从一根树枝荡跃到另一根树枝上。我的四肢很长，能牢牢地抓住树枝！

凭借灵活的四肢和弯曲成钩的手掌，长臂猿可以轻松地勾住树枝，在树林间荡跃前进。

我的哥哥更厉害！它可以荡跃着前进，穿越整片森林。

对长臂猿来说，单手把身体悬吊起来，简直小菜一碟！

看，我的手臂可以伸这么长！

长臂猿小知识

- 长臂猿喜欢坐着睡觉。
- 长臂猿非常擅长在地面或藤蔓上行走，走路的样子就像人类一样。
- 长臂猿爱吃水果、树叶和昆虫等，比如香甜的无花果令它们欲罢不能！

长臂猿宝宝
（A baby gibbon）

我出生在热带雨林（rainforest）中高大的树上，和爸爸妈妈、兄弟姐妹生活在相亲相爱的大家庭里。这些高高的树木，就是我们活动的乐园。

我什么时候才能长大呀？我都等不及要去树林中"荡秋千"啦！

1岁左右，长臂猿幼崽开始学习在树林间腾空荡跃。

香甜的梦

夜幕降临,大猩猩会用树叶和树枝搭建舒适的巢。第二天醒来(wake up),它们就能在床上享用美味的树叶早餐(breakfast)啦!

玩累了,我就和兄弟姐妹一起休息。

我10个月大了
（I'm ten months old）

现在，我终于可以自己走路啦！不过我还小呢，走一段路就会觉得累（tired）。这时，我会爬到哥哥身上，让它背着我，搭"顺风车"。大部分时间，我都和兄弟姐妹形影不离。

大猩猩喜欢在中午小睡一会儿。

我们快到目的地了吗?

大猩猩的一天

大猩猩通常会在清晨和傍晚觅食、进食。其他时间呢?那就睡一睡、玩一玩再梳理梳理(grooming)毛发喽!

我是一只大猩猩
（I'm a gorilla）

我和家人生活在茂密的森林（forest）。我现在才4个月大，还不能自己行走（walk），当然得紧紧抱着妈妈了。妈妈外出觅食时，我就紧紧地抱住妈妈并抓住它的皮毛（fur）。

免费搭乘

在出生后的约5个月的时间里，我都只能一直紧紧地抱着妈妈并抓住它的皮毛，和它一起行动。之后，我日渐强壮，就可以骑在妈妈的背（back）上或肩膀（shoulder）上了。

猩猩小知识

- 一只成年红毛猩猩大约有8个成年人那么强壮。
- 成年红毛猩猩的手臂伸展开的长度可达2.5米。
- 红毛猩猩会使用多种工具。有的猩猩会把树叶当成纸巾，用来擦干净脸上的食物残渣。

弟弟，小心点儿！

家庭生活

红毛猩猩幼崽和妈妈一起生活到8岁左右，之后它们就离开（go off）妈妈，独立生活。

学习爬树
（Learning to climb）

我现在3岁啦！哥哥姐姐正在教我爬树（climb tree）的本领，不过，我还是小心翼翼地，尽量离妈妈近一些。

红毛猩猩的四肢十分强壮有力。

嘿，快看！我会爬树啦！

红毛猩猩的牙齿非常适合啃食水果。

3岁左右,我们必须要学会攀爬。

我们荡来荡去,就像荡秋千一样。

水果早餐

看!这只红毛猩猩正在吃一种热带水果——榴莲(durian)。红毛猩猩爱吃水果,不过它们也吃树叶(leaf)、树皮、花朵(flower)和昆虫(insect)等。

我是一只红毛猩猩
（I'm an orang-utan）

我生活在枝繁叶茂的热带丛林（jungle），和妈妈一起住在高大的树上。妈妈经常会抓着树枝从一棵树荡到另一棵树上，这时，我会紧紧地抱住（hang on）它。

抓紧了！

妈妈会一直养育红毛猩猩幼崽到8岁左右，之后幼崽就要独立生活了。

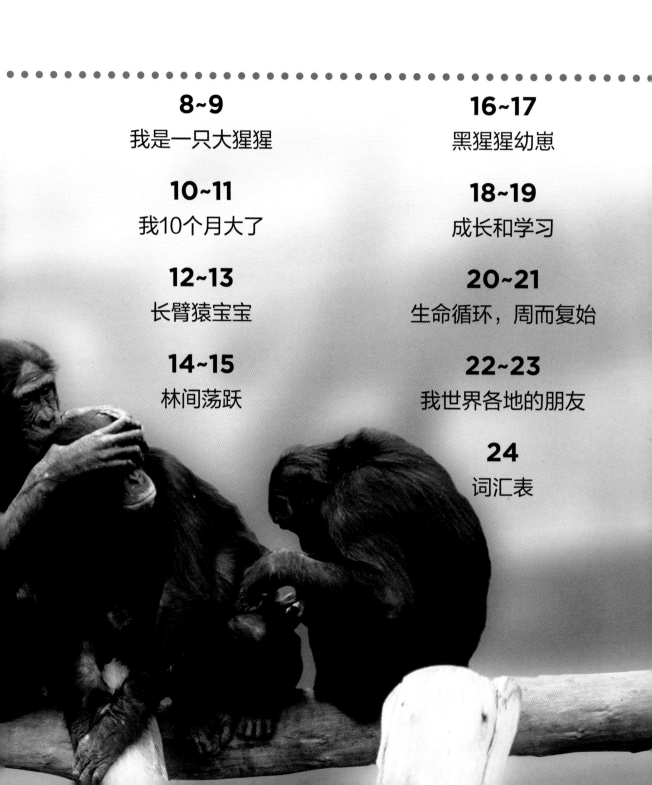

8~9
我是一只大猩猩

10~11
我10个月大了

12~13
长臂猿宝宝

14~15
林间荡跃

16~17
黑猩猩幼崽

18~19
成长和学习

20~21
生命循环，周而复始

22~23
我世界各地的朋友

24
词汇表

Original Title: Apes
Copyright © Dorling Kindersley Limited, 2006
A Penguin Random House Company

本书简体中文版授权由人民邮电出版社独家出版，仅限于中国境内（不包括香港、澳门、台湾地区）销售。未经出版者书面许可，不得以任何方式复制或发行本书中的任何部分。

图书在版编目（CIP）数据

DK动物成长奥秘：看！我在长大：汉英对照 / 英国DK公司编；严景熙等译. — 北京：人民邮电出版社，2022.3
ISBN 978-7-115-57207-3

Ⅰ.①D… Ⅱ.①英… ②严… Ⅲ.①动物—儿童读物—汉、英 Ⅳ.①Q95-49

中国版本图书馆CIP数据核字(2021)第218519号

内 容 提 要

本书以孩子们熟悉的超过13种动物（猿、熊、蝴蝶、鸭子、大象、农场动物、青蛙、小猫、大熊猫、企鹅、小狗、兔子、海龟）为主题，以中英双语版图书的形式，分别详细地介绍了每种动物从繁殖到成长各个生命阶段的形态和习性等。书中独特新奇的角度、色彩绚烂的图片、活泼生动的文字，可以让孩子们在了解生命成长奥秘的同时，还能练习英语阅读。

◆ 编　　　英国DK公司
　 译　　　严景熙 李梓轩 Jeremy Wang 汪祝铭 等
　 责任编辑　宁 茜
　 责任印制　陈 犇
◆ 人民邮电出版社出版发行　北京市丰台区成寿寺路11号
　 邮编 100164　电子邮件 315@ptpress.com.cn
　 网址 https://www.ptpress.com.cn
　 广东金宜发包装科技有限公司印刷
◆ 开本：787×935　1/16
　 印张：39　　　　　　　2022年3月第1版
　 字数：620千字　　　　 2022年3月广东第1次印刷
　 著作权合同登记号　图字：01-2021-2060号

总定价：238.00元（全13册）
读者服务热线：(010)81055493　印装质量热线：(010)81055316
反盗版热线：(010)81055315
广告经营许可证：京东市监广登字20170147号

For the curious
www.dk.com

目录 Contents

4~5
我是一只红毛猩猩

6~7
学习爬树

DK 动物成长奥秘
看！我在长大（中英双语版）

猿

英国 DK 公司◎编

严景熙 李梓轩 Jeremy Wang 汪祝铭◎译

鹰之舞 沈成◎审

人民邮电出版社

北京

DK WATCH ME GROW
BEAR

Contents

4~5
I'm a bear

6~7
Dad fights for mum

8~9
I'm born in the den

10~11
I'm taking my first steps

12~13
A walk in the woods

14~15
Mum on the lookout

16~17
I'm learning to fish

18~19
Time for a long sleep

20~21
The circle of life

22~23
My friends from around the world

24
Glossary

Come follow us and see how we GROW!

I'm a bear

I am the largest animal in the woods and forests where I live. I can swim and climb trees. My thick fur keeps me warm in winter.

I am the king of the forest.

Peering around

Bears are curious. They often stand up on their back legs to look over the bushes.

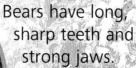

Bears have long, sharp teeth and strong jaws.

The bear's sharp claws help it to hunt for food.

Turn the page and watch me grow...

Dad fights for mum

Grrr Grrrrr

Dad had to fight with other male bears to see who would get to mate with mum. Dad was the biggest bear, so he won.

The younger and smaller bear will lose the fight.

Their fight may look scary, but these bears are not trying to hurt each other. They are wrestling to see who is the strongest.

Go away! I am the strongest.

When the cubs are born

Mum and dad only stay together for one or two weeks. Then dad will leave and mum will raise the cubs on her own.

Gripping facts

- Mum and dad mate in June or July, but their cubs won't be born until the next January.
- Each bear has its own area, called a territory.
- Female bears can start having cubs when they are about four or five years old.

I'm born in the den

It's winter when we are born, but mum has found a warm, snug place to protect us from the cold and snow. The place where we live is called our den.

Bald and blind
This week-old cub was born tiny, blind and almost completely bald. It depends on its mother for everything.

During the cold winter, the mother bear does not eat. She spends all her time feeding her cubs, keeping them warm, and sleeping.

After one month, our eyes are open and we have fur.

I'm taking my first steps

After two months in the den, the weather is warmer and it's time for us to start exploring the world. We are ready to start learning all about how to be bears.

Watch and learn
Bear cubs are curious. They learn by watching mum and by playing.

My white collar makes it easy for mum to find me.

Milk from mum
Bear cubs get most of their food by drinking milk from their mother.

A walk in the woods

We are three months old and very good at walking. It's time for mum to teach us about the different foods we can find in the woods. We stay close to mum so she can protect us.

Berry nice
The young cubs feed on berries, insects and small animals such as frogs... yummy!

Wait for me, I'm right behind you!

The cubs have many new smells to learn about.

Mum is on the lookout for male bears who might hurt her cubs.

Bear-faced facts

- Bears have a great sense of smell. They can smell food up to 2 km (1.2 m) away.
- The cubs drink milk from mum until they are about five months old.
- By watching mum carefully, the cubs learn what foods are safe to eat.

Mum on the lookout

Mum keeps watch for danger while we practise climbing trees. It's not as easy as it looks.

When she stands up, mum can see a long way away.

Up we go...

...Whoops

Hold on tight!

Danger alert
A mother bear will chase off almost any danger. But hungry wolves are on the lookout for cubs that get separated from their mum.

Wolves

I'm learning to fish

It's time for me to join the older bears for my first fishing lesson. I learn how to catch the slippery fish by watching and copying the other bears.

This six-month-old cub is too small to stand in the fast-moving water, so he fishes from the shallow water at the edge.

The bears must wait patiently for the salmon to leap out of the water.

Clams for lunch

Bears love seafood of all kinds. This mother and cub are digging for clams. Their long claws are perfect clam openers.

Time for a long sleep

Winter is here, and there is not much food to eat. It's time to find a cosy place to go to sleep. We will sleep until spring, when there is more to eat.

These are my footprints

The bears have grown fat during summer and autumn. The extra weight will nourish them until spring.

Bears grow a thick coat of winter fur.

These bears have dug a snow cave to spend the winter in.

I'm tired...time for bed.

The cubs will share their mother's den until they are almost two.

The circle of life goes round and round

Now you know how I turned into a grown-up bear.

My friends from around the world

The Sloth Bear lives in forests in India and Nepal. Its favourite food is termites.

The Giant Panda lives in China and eats only bamboo leaves and shoots.

These American Black Bears live in forests and don't like open spaces.

I'm the world's biggest bear,

My bear friends come in all sorts of different shapes and sizes.

Spectacled Bears are from South America. They build their nests in trees.

The smallest bear is the Asian Sun Bear.

and I live in the cold, snowy Arctic.

I'm a Black Bear cub.

Polar Bears' favourite foods are seals and walruses.

Friendly facts

- Polar Bears are very good swimmers. They can swim for long distances without a rest.

- A female bear is called a sow. A male bear is called a boar.

- Bears see in colour and have much better hearing and smell than humans.

Glossary

Collar
The ring of white fur around the neck of all young cubs.

Nursing
When the mother bear feeds the cubs with her milk.

Hibernation
Bears usually sleep all winter. This is called hibernation.

Salmon
A type of fish that is very fatty. It is good food for bears.

Cub
For the first year of its life, a baby bear is called a cub.

Omnivore
An animal that eats almost anything. Bears are omnivores.

Acknowledgements
感谢以下人员及机构提供图片：

Position key: c=centre; b=bottom; l=left; r=right; t=top.
1: Lynn Rogers; 2-3: Nature Picture Library/Tom Mangelsen; 4: Still Pictures/Klein/Hubert; 5: Corbis/Greg Probst (r), /Michael T. Sedam (cl), /Steve Kaufman (bl); Oxford Scientific Films/Mathias Breiter (tl); Lynn Rogers (br); 6-7: ImageState/Pictor; 7: Michael S. Quinton (tr); 8: FLPA - Images of Nature/L. LeeRue (cl); 8-9: Lloyd Beebe; 10: Ardea London Ltd (cl main); 10: McDonald Wildlife Photography (cl, youngest cub); 10-11: McDonald Wildlife Photography; 11: Corbis/Kennan Ward (cr); 12: Nature Picture Library/Tom Mangelsen (l); 12-13: Nature Picture Library/ Steffan Widstrand; 14: Nature Picture Library/Steffan Widstrand (c); 14-15: McDonald Wildlife Photography (c), Getty Images/J.P. Fruchet (main); 15: McDonald Wildlife Photography (c, cr); 16: ImageState/Pictor (r), McDonald Wildlife Photography (bl); 17: ImageState/Pictor (tr, l), Lynn Rogers (cr); 18: Ardea London Ltd/Stefan Meyers (c), Corbis/Dan Guravich (tl), FLPA - Images of Nature/M. Newman (tr), Still Pictures/Francois Gilson (main); 19: Ardea London Ltd/Johan De Meester (b); 20: Corbis/Joe McDonald (cr below), Team Husar Wildlife Photography (tr), ImageState/Pictor (bc), FLPA - Images of Nature/ L. LeeRue (tc), McDonald Wildlife Photography (cr above), Lynn Rogers(tl, cl, bl, br), Andrew Rouse (c); 21: Corbis/Gunter Marx Photography (foreground); 22: Lynn Rogers (tl); 22-23: Oxford Scientific Films/Daniel Cox; 23: Oxford Scientific Films/Terry Heathcote (br); Lynn Rogers (cr); 24: Nature Picture Library/Tom Mangelsen (br), Lloyd Beebe (tr), ImageState/Pictor (cl), FLPA - Images of Nature/M. Hoshino/Minden (bl), McDonald Wildlife Photography (tl).

其他图片版权属于多林·金德斯利公司。欲了解更多信息请访问DK Images网站。

词汇表 Glossary

颈毛 Collar
熊宝宝脖子上的一圈白毛。

哺乳 Nursing
熊妈妈给熊宝宝喂母乳的过程。

冬眠 Hibernation
某些动物在冬季生命活动减少、体温下降和陷入昏睡的状态。

鲑鱼 Salmon
一种很肥美的鱼，熊特别喜欢捕食这种鱼。

幼崽 Cub
出生不满一年的熊宝宝被称为幼崽。

杂食性动物 Omnivore
既吃植物也吃动物的动物，熊就是一种杂食性动物。

致谢 Acknowledgements

感谢以下人员及机构提供图片：

Position key: c=centre; b=bottom; l=left; r=right; t=top.
1: Lynn Rogers; 2-3: Nature Picture Library/Tom Mangelsen; 4: Still Pictures/Klein/Hubert; 5: Corbis/Greg Probst (r), /Michael T. Sedam (cl), / Steve Kaufman (bl); Oxford Scientific Films/Mathias Breiter (tl); Lynn Rogers (br); 6-7: ImageState/Pictor; 7: Michael S. Quinton (tr); 8: FLPA - Images of Nature/L. LeeRue (cl); 8-9: Lloyd Beebe; 10: Ardea London Ltd (cl main); 10: McDonald Wildlife Photography (cl, youngest cub); 10-11: McDonald Wildlife Photography; 11: Corbis/Kennan Ward (cr); 12: Nature Picture Library/Tom Mangelsen (l); 12-13: Nature Picture Library/ Steffan Widstrand; 14: Nature Picture Library/Steffan Widstrand (c); 14-15: McDonald Wildlife Photography (c), Getty Images/J.P. Fruchet (main); 15: McDonald Wildlife Photography (c, cr); 16: ImageState/Pictor (r), McDonald Wildlife Photography (bl); 17: ImageState/Pictor (tr, l), Lynn Rogers (cr); 18: Ardea London Ltd/Stefan Meyers (c), Corbis/Dan Guravich (tl), FLPA - Images of Nature/M. Newman (tr), Still Pictures/Francois Gilson (main); 19: Ardea London Ltd/Johan De Meester (b); 20: Corbis/Joe McDonald (cr below), Team Husar Wildlife Photography (tr), ImageState/Pictor (bc), FLPA - Images of Nature/ L. LeeRue (tc), McDonald Wildlife Photography (cr above), Lynn Rogers(tl, cl, bl, br), Andrew Rouse (c); 21: Corbis/Gunter Marx Photography (foreground); 22: Lynn Rogers (tl); 22-23: Oxford Scientific Films/Daniel Cox; 23: Oxford Scientific Films/Terry Heathcote (br); Lynn Rogers (cr); 24: Nature Picture Library/Tom Mangelsen (br), Lloyd Beebe (tr), ImageState/Pictor (cl), FLPA - Images of Nature/M. Hoshino/Minden (bl), McDonald Wildlife Photography (tl).

其他图片版权属于多林·金德斯利公司。欲了解更多信息请访问DK Images网站。

我世界各地的朋友（friend）长相不同，体形各异。

眼镜熊（Spectacled Bear）来自南美洲，它们在树上安家。

马来熊（Asian Sun Bear）是世界上最小的熊。

生活在严寒难耐、冰雪覆盖的北极。

我是一只黑熊宝宝。

北极熊（Polar Bear）最喜欢的食物是海豹（seal）和海象（walrus）。

熊的小知识

- 北极熊是游泳能手，一口气可以游很长一段距离。
- 母熊在英语中被称为"sow"，公熊被称为"boar"。
- 熊能够辨别颜色，它们的听觉和嗅觉远远超过人类。

我世界各地的朋友
My friends from around the world

大熊猫（Giant Panda）生活在中国（China），它喜欢吃竹叶和竹笋。

懒熊（Sloth Bear）生活在印度和尼泊尔的森林里。它们最爱吃白蚁（termite）。

美洲黑熊（American Black Bear）生活在森林里，不喜欢开阔地带。

我是世界上最大的熊，

再见啦,我要去睡觉了。

生命循环，周而复始
The circle of life goes round and round

现在你知道我怎样长成一只成年熊了吧！

Now you know how I turned into a grown-up bear.

熊已经挖好了雪洞，准备过冬。

好累啊，我要去睡觉了。

熊宝宝大约在两岁之前都会和熊妈妈在一起生活。

冬眠
(Time for a long sleep)

冬季来临,食物变少了,该找个暖和的地方(place)冬眠了。我们到来年春天(spring)才会结束冬眠,那时就不愁没食物了。

这是我的脚印。

熊在夏季和秋季(summer and autumn)这两个季节会长胖。脂肪的储存可以帮助它们在冬季维持生命,直到春季来临。

熊在冬季会换上一层比在其他季节更厚的毛皮"外套"(coat)。

熊必须耐心等待着鲑鱼（salmon）跃出水面，然后果断出击，一击而中。

蛤蜊盛宴

熊喜欢吃各种海鲜（seafood）。这位熊妈妈正带着宝宝挖蛤蜊（clam）。它们的爪子很长，用来开蛤蜊再合适不过了。

学习捕鱼
(I'm learning to fish)

现在是时候加入前辈们,开始上第一堂捕鱼课（lesson）啦!我先观察它们是怎样做的,然后学着它们的样子,向滑溜溜的（slippery）鱼发起进攻。

水流太急了,这只6个月大（six-month-old）的熊宝宝在水中站不稳,所以它只能在岸边水浅的地方捕鱼。

哎呀,我要掉下去了!

快抓紧!

危险警报

一旦察觉到危险(danger)靠近,熊妈妈会拼尽全力保护好熊宝宝。可是,饥肠辘辘的狼(wolf)依然紧紧盯着它们,专门等着与熊妈妈走散的熊宝宝。

狼

熊妈妈站岗
(Mum on the lookout)

在我们练习爬树（climbing tree）时，妈妈就在一旁站岗，保证我们在安全的环境里学习本领。站岗这活儿可不像看起来那么简单哦！

当熊妈妈用后腿站立（stand up）时，它能看得更远。

向上爬呀……

熊的小知识

- 熊的嗅觉十分灵敏,2千米以外的气味都能闻到。
- 在出生后约5个月的时间里,熊宝宝一直喝母乳。
- 通过细心观察妈妈的行为,熊宝宝能学会辨别哪些食物可以食用。

熊妈妈正在站岗放哨,以提防有公熊(male bear)突然袭击熊宝宝。

林中漫步
(A walk in the woods)

我们3个月大（three months old）了，走路（walk）已经很稳了，是时候让妈妈教我们如何在森林里寻找美食了。我们紧跟在妈妈身后，这样妈妈就能随时保护（protect）我们了。

浆果真美味！

熊宝宝爱吃浆果（berry）、昆虫（insect）和某些小型动物，比如青蛙等。
森林里可以找到的美食可真多呀！

等等我，我掉队啦！

熊宝宝还要学会识别各种气味。

我可是真正的"白领",这么显眼的打扮让妈妈一眼就能找到我。

母乳喂养

熊宝宝通过喝母乳来获得成长所需的大部分营养。

迈出第一步
(I'm taking my first steps)

在洞穴中生活了两个月,天气（weather）渐渐暖和起来,我们终于可以外出探索世界（explore the world）啦！我们做好了准备（ready）,开始学习如何成为一只优秀的熊。

观察和学习

熊宝宝的好奇心很强,它们通过玩耍（play）和观察妈妈的行为来学习生存的本领。

在寒冷的冬天,熊妈妈不进食(eat)。它所有的时间都用来喂养和温暖熊宝宝,以及冬眠。

一个月后,我们可以睁开双眼了,身上的毛也长出来了。

生于洞穴
(I'm born in the den)

我们出生时正是冬天,但是妈妈找到了一个既暖和(warm)又舒适(snug)的洞穴供我们居住,免受风雪(snow)的侵袭,这样我们就可以度过这个寒冷的冬天了。

生命之初

一周大的熊宝宝个头小小的(tiny),并且什么也看不见,全身几乎没有毛(bald)。生活上的一切(everything)都要依赖妈妈。

走开！我才是最强壮的！

我们出生啦！

熊宝宝出生后，熊妈妈和熊爸爸只会朝夕相处一到两周（one or two weeks）。之后，熊爸爸就离开了，留下熊妈妈独自抚养熊宝宝。

熊的小知识

- 熊爸爸和熊妈妈在每年的六七月交配，而熊宝宝到来年一月才会降生。

- 熊的领地意识很强，每只熊都有自己的地盘。

- 母熊大约长到四五岁就可以生宝宝了。

为熊妈妈而战
(Dad fights for mum)

我的爸爸必须战胜其他公熊，才能赢得与妈妈交配（mate）的权力。爸爸的块头最大，所以它赢（win）啦！

啵啵…… 呜啵啵……

那些年幼或体弱的公熊在搏斗（fight）中往往会败下阵来。

公熊的搏斗场面看起来有点吓人（scary），但其实它们并不想伤害对方（hurt each other），摔跤（wrest）只是为了证明自己更强壮。

东瞧瞧，西看看

熊有很强的好奇心。它们常常用后腿站立，以便视线越过灌木丛（bush）观察远方。

熊的牙齿（teeth）又长又尖，下颌（jaw）非常有力。

翻开下一页，看看我是怎样长大的吧！

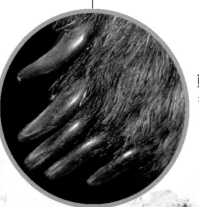

熊的爪子（claw）很锋利，能够帮助它们猎捕食物（food）。

我是一只熊
(I'm a bear)

我是森林（forest）中体形最大的动物（animal），擅长游泳和爬树。我长着一身厚厚的皮毛，冬天（winter）再冷也能保暖。

我是森林之王。

14~15
熊妈妈站岗

16~17
学习捕鱼

18~19
冬眠

20~21
生命循环，周而复始

22~23
我世界各地的朋友

24
词汇表

跟随我的脚步，看着我成长吧！

目录 Contents

Original Title: Bear
Copyright © Dorling Kindersley Limited, 2003
A Penguin Random House Company

本书简体中文版授权由人民邮电出版社独家出版，仅限于中国境内（不包括香港、澳门、台湾地区）销售。未经出版者书面许可，不得以任何方式复制或发行本书中的任何部分。

4~5
我是一只熊

6~7
为熊妈妈而战

8~9
生于洞穴

10~11
迈出第一步

12~13
林中漫步

For the curious
www.dk.com

DK 动物成长奥秘
看！我在长大（中英双语版）

熊

英国 DK 公司◎编
邹思苇 陈楚霖 谢若宸 赵泓羽◎译
鹰之舞 沈成◎审

人民邮电出版社
北京

WATCH ME GROW
BUTTERFLY

Contents

watch us GROW!

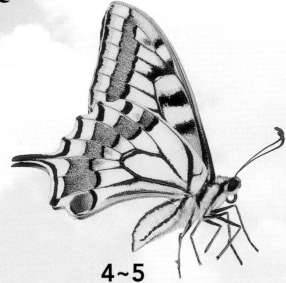

4~5
I'm a butterfly

6~7
Before I was born

8~9
Time to hatch

10~11
I am growing bigger

12~13
I'm very hungry

14~15
Holding on tight

16~17
Time to break out

18~19
Get ready to fly

20~21
The circle of life

22~23
My friends from around the world

24
Glossary

I'm a butterfly

I use my colourful wings to fly from flower to flower, and drink nectar through my long, curly tongue.

Its body is covered with millions of soft hairs.

Antennae help the butterfly to smell and to balance.

The butterfly sucks nectar through its hollow proboscis, which uncurls to act like a straw!

This is a close-up of my scales.

A butterfly's wings are made up of thousands of tiny scales.

Now turn the page and watch me GROW...

Before I was born

Mum and dad met while they were flying in a field. They flew around each other for a few minutes, and then landed on a flower to mate.

After mating, the male flies away and the female looks for a plant where she can lay her eggs.

Egg laying

The female curves her bendy body towards a leaf to lay her eggs. The eggs are sticky so they don't roll away.

Home sweet home

Each type of butterfly will lay its eggs on only a few plants. The type of butterfly in this book likes carrot and fennel plants best.

Giant fennel

Wild carrot

Time to hatch

After about five days of growing inside my egg, I am ready to hatch out as a tiny caterpillar. I have to chew my way out of my egg. It's hard work.

This two-day-old egg will soon start to change colour.

It takes many hours for the caterpillar to chew its way out of the egg.

Home sweet home

Butterflies live just about anywhere there are flowers. Spring is the best time to find their tiny eggs, but you have to look very carefully.

My eggshell is my first meal!

I am growing bigger

The more I eat, the bigger I get. Soon, I can't fit into my skin any more. It's time for me to shed my old skin and grow a bigger skin. Each skin is a different colour.

7 days   12 days   18 days

Danger alert
When the caterpillar senses danger, this orange scent horn pops up and gives off a stinky odour to scare away enemies.

Sometimes the caterpillar eats its old skin after shedding.

Caterpillars do not have lungs. Instead, they breathe through holes in their skin.

I'm very hungry

I am now three weeks old and I have to eat all the time. I have only a few weeks to store enough energy to change into a butterfly.

These are the caterpillar's teeth!

I don't sleep, I just eat, and eat, and eat.

Sharp spikes warn other animals to stay away.

Munch, crunch!

Caterpillars have lots of feet so they can grip on to branches.

Crunchy facts

- Caterpillars are fussy eaters. Most eat only one or two kinds of plant.
- If you grew as fast as a caterpillar, you would be as big as a truck in two weeks!

Holding on tight

After about four weeks I find a nice, strong branch and spin some sticky silk thread to help me hang on. Now I'm ready to shed my skin for the last time.

The pad of coiled thread on the tail is called the pillow.

It's time to begin the change into a butterfly.

The thread around the caterpillar's middle is called the belt.

The caterpillar sheds its skin. Underneath, a shell has formed.

The shell will harden into a protective case called a chrysalis.

Inside, the caterpillar turns into a lump of soft, squidgy jelly.

Time to break out

It's been almost three weeks since I started changing. The soft jelly inside my chrysalis is turning into the body of a beautiful butterfly.

See-through package
When it is time to hatch, the chrysalis turns clear. Look closely. Can you see the colour of the new butterfly?

I push and I shove and my chrysalis splits open.

When the butterfly emerges, its wings are wet and crumpled.

The butterfly pumps blood into its wings to help them expand.

Hatching facts

- Some butterflies spend the winter in their chrysalis, then hatch in the spring.
- A butterfly's skeleton is on the outside of its body to protect it.

Get ready to fly

It only takes a few minutes for my wings to dry off. Now I am ready to look for flowers, which is where I will find my new food.

The empty chrysalis is left behind.

My wings are dry and I'm ready to fly.

The adult butterfly will live for about a month, so there is not much time to find a mate.

Slurp slurp

From now on, the butterfly will drink nectar through its tongue, or proboscis.

The circle of life goes round and round

Now you know how I turned into a beautiful butterfly.

My friends from around the world

To scare off birds, the Peacock Butterfly has spots that look like eyes.

This Leaf Butterfly hides by looking like a dead leaf.

The Clubtail Butterfly lives in warm, wet rainforests.

Can you find me?

I'm a Pygmy Butterfly and I'm the smallest!

I'm the biggest. This pale green

My butterfly friends around the world come in all the colours of the rainbow.

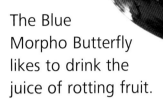

The Blue Morpho Butterfly likes to drink the juice of rotting fruit.

This butterfly from South America is called the 88 Butterfly. Can you see why?

The Malachite Butterfly eats bird droppings!

The Queen Alexandra's Birdwing comes from New Guinea.

shape shows my real size.

Butterfly facts

- Monarch Butterflies travel 8,800 km (4,000 miles) each year, from the Great Lakes to the Gulf of Mexico and back.

- There are about 28,000 different types of butterflies.

- A butterfly cannot fly if its body temperature falls below 30°C (86°F).

Glossary

Proboscis
The butterfly uses this part of its body to drink nectar.

Shedding
The caterpillar loses its old skin and grows bigger skin.

Hatching
When the baby caterpillar first comes out of its egg.

Chrysalis
The stage when the caterpillar is changing into a butterfly.

Caterpillar
The second stage of a butterfly's life cycle, after egg.

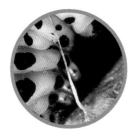

Silk
The thread the caterpillar makes to hold it onto a branch.

Acknowledgements
感谢以下人员及机构提供图片:

Jerry Young, Andy Crawford, Frank Greenaway, Colin Keates, Natural History Museum, Derek Hall, Eric Crichton, Kim Taylor, Jane Burton (Key: a=above; c=centre; b=below; l=left; r=right; t=top)
1: Alamy Images t; 2-3: N.H.P.A./Laurie Campbell b; 3: Oxford Scientific Films/Stan Osolinski tr; 4: Duncan McEwan/naturepl网站 clb; 4-5: Oxford Scientific Films/Raymond Blythe; 5: Science Photo Library/Andrew Syred tl; 6-7: Flowerphotos/Carol Sharp; 6: Richard Revels; 7: Corbis/George McCarthy (butterfly) ca; 7: Windrush Photos/Richard Revels (leaf & egg) ca; 8-9: FLPA - Images of nature/Ian Rose (background); 8-9: Windrush Photos/Richard Revels c; 9: Woodfall Wild Images/Richard Revels b; 10: Ardea London Ltd/Pascal Goetgheluck clb;

10-11: Oxford Scientific Films/Raymond Blythe; 12-13: FLPA - Images of Nature/Ian Rose (background); 12-13: Hans Christoph Kappel/naturepl网站 (caterpillar); 12: N.H.P.A./Daniel Heuclin cra; 14-15: Richard Revels (caterpillar); 15: Richard Revels tr, cr, br; 16: Richard Revels cla, bc; 17: Richard Revels; 18: Richard Revels; 19 Ingo Arndt/naturepl网站 r; 20: FLPA - Images of nature/Roger Wilmshurst c; 20: Hans Christoph Kappel/naturepl网站 tl; 20: Richard Revels cla, clb, bcl; 21: Sonia Halliday Photographs/Sister Daniel (background); 21: Hans Christoph/naturepl网站 c; 24: Ardea London Ltd/Ian Beames br; 24: Oxford Scientific Films/Raymond Blythe tr.

其他图片版权属于多林·金德斯利公司。欲了解更多信息请访问DK Images网站。

My butterfly friends around the world come in all the colours of the rainbow.

This butterfly from South America is called the 88 Butterfly. Can you see why?

The Blue Morpho Butterfly likes to drink the juice of rotting fruit.

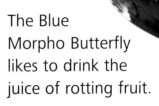

The Malachite Butterfly eats bird droppings!

The Queen Alexandra's Birdwing comes from New Guinea.

shape shows my real size.

Butterfly facts

- Monarch Butterflies travel 8,800 km (4,000 miles) each year, from the Great Lakes to the Gulf of Mexico and back.
- There are about 28,000 different types of butterflies.
- A butterfly cannot fly if its body temperature falls below 30°C (86°F).

Glossary

Proboscis
The butterfly uses this part of its body to drink nectar.

Shedding
The caterpillar loses its old skin and grows bigger skin.

Hatching
When the baby caterpillar first comes out of its egg.

Chrysalis
The stage when the caterpillar is changing into a butterfly.

Caterpillar
The second stage of a butterfly's life cycle, after egg.

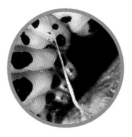

Silk
The thread the caterpillar makes to hold it onto a branch.

Acknowledgements
感谢以下人员及机构提供图片：

Jerry Young, Andy Crawford, Frank Greenaway, Colin Keates, Natural History Museum, Derek Hall, Eric Crichton, Kim Taylor, Jane Burton (Key: a=above; c=centre; b=below; l=left; r=right; t=top)
1: Alamy Images t; 2-3: N.H.P.A./Laurie Campbell b; 3: Oxford Scientific Films/Stan Osolinski tr; 4: Duncan McEwan/naturepl网站 clb; 4-5: Oxford Scientific Films/Raymond Blythe; 5: Science Photo Library/Andrew Syred tl; 6-7: Flowerphotos/Carol Sharp; 6: Richard Revels; 7: Corbis/George McCarthy (butterfly) ca; 7: Windrush Photos/Richard Revels (leaf & egg) ca; 8-9: FLPA - Images of nature/Ian Rose (background); 8-9: Windrush Photos/Richard Revels c; 9: Woodfall Wild Images/Richard Revels b; 10: Ardea London Ltd/Pascal Goetgheluck clb; 10-11: Oxford Scientific Films/Raymond Blythe; 12-13: FLPA - Images of Nature/Ian Rose (background); 12-13: Hans Christoph Kappel/naturepl网站 (caterpillar); 12: N.H.P.A./Daniel Heuclin cra; 14-15: Richard Revels (caterpillar); 15: Richard Revels tr, cr, br; 16: Richard Revels cla, bc; 17: Richard Revels; 18: Richard Revels; 19 Ingo Arndt/naturepl网站 r; 20: FLPA - Images of nature/Roger Wilmshurst c; 20: Hans Christoph Kappel/naturepl网站 tl; 20: Richard Revels cla, clb, bcl; 21: Sonia Halliday Photographs/Sister Daniel (background); 21: Hans Christoph/naturepl网站 c; 24: Ardea London Ltd/Ian Beames br; 24: Oxford Scientific Films/Raymond Blythe tr.

其他图片版权属于多林・金德斯利公司。欲了解更多信息请访问DK Images网站。

我世界各地的蝴蝶朋友身披如彩虹（rainbow）般绚烂的颜色。

这是来自南美洲的88蝶（88 Butterfly，学名红涡蛱蝶）。你知道它为什么叫这个名字吗？

蓝闪蝶（Blue Morpho Butterfly）喜欢吸食腐烂的水果。

绿帝蛱蝶竟然以鸟粪为食。

亚历山大女皇鸟翼凤蝶（Queen Alexandra's Birdwing）来自新几内亚。

绿色的轮廓才是我真实的大小！

蝴蝶小知识

🦋 黑脉金斑蝶每年从五大湖区迁徙到墨西哥湾，往返路程约8800千米。

🦋 世界上约有28,000种蝴蝶。

🦋 如果蝴蝶的体温低于30℃，它就无法飞行了。

词汇表 Glossary

口器
Proboscis
蝴蝶用来吸食花蜜的器官。

蜕皮
Shedding
旧的表皮脱落，长出更大的新皮的过程。

孵化
Hatching
毛毛虫宝宝初从卵里钻出来的过程。

蛹
Chrysalis
毛毛虫化成蝴蝶的中间阶段。

毛毛虫
Caterpillar
蝴蝶生命周期的第二个阶段（在卵之后）。

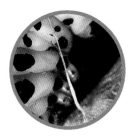
丝
Silk
毛毛虫为了抓住树枝而吐出来的线状物质。

致谢 Acknowledgements

感谢以下人员及机构提供图片：

Jerry Young, Andy Crawford, Frank Greenaway, Colin Keates, Natural History Museum, Derek Hall, Eric Crichton, Kim Taylor, Jane Burton (Key: a=above; c=centre; b=below; l=left; r=right; t=top)
1: Alamy Images t; 2-3: N.H.P.A./Laurie Campbell b; 3: Oxford Scientific Films/Stan Osolinski tr; 4: Duncan McEwan/naturepl.com clb; 4-5: Oxford Scientific Films/Raymond Blythe; 5: Science Photo Library/Andrew Syred tl; 6-7: Flowerphotos/Carol Sharp; 6: Richard Revels; 7: Corbis/George McCarthy (butterfly) ca; 7: Windrush Photos/Richard Revels (leaf & egg) ca; 8-9: FLPA - Images of nature/Ian Rose (background); 8-9: Windrush Photos/Richard Revels c; 9: Woodfall Wild Images/Richard Revels b; 10: Ardea London Ltd/Pascal Goetgheluck clb; 10-11: Oxford Scientific Films/Raymond Blythe; 12-13: FLPA - Images of Nature/Ian Rose (background); 12-13: Hans Christoph Kappel/naturepl.com (caterpillar); 12: N.H.P.A./Daniel Heuclin cra; 14-15: Richard Revels (caterpillar); 15: Richard Revels tr, cr, br; 16: Richard Revels cla, bc; 17: Richard Revels; 18: Richard Revels; 19 Ingo Arndt/naturepl.com r; 20: FLPA - Images of nature/Roger Wilmshurst c; 20: Hans Christoph Kappel/naturepl.com tl; 20: Richard Revels cla, clb, bcl; 21: Sonia Halliday Photographs/Sister Daniel (background); 21: Hans Christoph/naturepl.com c; 24: Ardea London Ltd/Ian Beames br; 24: Oxford Scientific Films/Raymond Blythe tr.

其他图片版权属于多林·金德斯利公司。欲了解更多信息请访问DK Images网站。

我世界各地的朋友
My friends from around the world

孔雀蛱蝶（Peacock Butterfly）身上长有像眼睛一样的斑点，用来吓走想捕食它的小鸟。

枯叶蝶（Leaf Butterfly）善于伪装成一片枯叶以隐藏自己。

锤尾凤蝶（Clubtail Butterfly）生活在温暖潮湿的热带雨林中。

你能找到我吗？我是小灰蝶，是世界上最小的蝴蝶！

我是世界上最大的蝴蝶，这层浅

生命循环，周而复始
The circle of life goes round and round

现在你知道我怎样长成一只美丽的蝴蝶了吧！

Now you know how I turned into a beautiful butterfly.

成年（adult）蝴蝶的寿命约为一个月，所以它们要抓紧时间寻找配偶。

享受美食

成年蝴蝶会用它的口器吸食花蜜，以填饱肚子。

准备振翅而飞吧
(Get ready to fly)

我的翅膀只需要几分钟就能干透（dry off）。我准备振翅飞向花丛啦，那里有很多我没有吃过的美食（food）！

空空的（empty）蛹壳被留在原地。

我的翅膀已经干透，可以随时起飞啦！

蝴蝶刚出来时，翅膀又湿（wet）又皱（crumpled）。

蝴蝶会将血液（blood）泵入翅膀，帮助翅膀展开（expand）。

蝴蝶小知识

- 有些蝴蝶会在蛹壳里过冬，直到春天才会破蛹而出。
- 为了保护自己的身体，蝴蝶的骨骼长在肌肉外面。

是时候破蛹而出啦
(Time to break out)

我变成蛹已经将近3周了。蛹壳里的"软果冻"(the soft jelly)正慢慢变成一只美丽的(beautiful)蝴蝶。

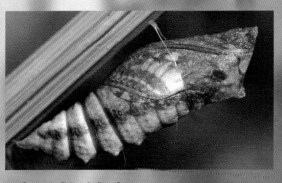

透明的外衣

快化蝶时,蛹壳会变得越来越透明(clear)。仔细观察(look closely),你能看到里面这只新生蝴蝶的颜色吗?

我推呀推、挤呀挤,终于把蛹壳挤破了。

是时候蜕变成蝴蝶了。

这条缠绕着毛毛虫的丝叫作保护带（the belt）。

当毛毛虫完成最后一次蜕皮时，新的外壳已经在旧皮肤下形成了。

这个外壳逐渐硬化成一个叫作蛹壳的保护套。

在蛹壳里，毛毛虫会软作一团，看起来就像果冻一样。

抓紧了
(Holding on tight)

大约4周后,我找到一根结实的树枝,开始在上面吐丝(spin)。这些黏黏的丝(silk)可以帮助我安稳地挂(hang)在树枝上。现在,我已经为最后一次蜕皮做好了充足的准备!

这个缠绕在毛毛虫尾部(tail)的丝团叫作丝垫(the pillow)。

毛毛虫身上尖锐的（sharp）刺（spike）可以警告（warn）其他动物，不要靠近。

嘎吱，嘎吱！

毛毛虫有很多只脚（feet），它们能紧紧抓住树枝（branch）。

毛毛虫小知识

- 毛毛虫很挑食。它们大多只吃一两种植物。

- 如果你长得像毛毛虫一样快，那么两周后，你就会变得和卡车一样大！

13

我很饿
(I'm very hungry)

3周大的我必须从早吃到晚。因为只有这样我才能在之后的几周内,储存(store)足够的能量(energy),变成一只美丽的蝴蝶。

这是毛毛虫的牙齿(teeth)!

我不睡觉,只专心吃呀吃,吃呀吃。

毛毛虫虽然没有肺（lung），但它可以通过皮肤上的气门来呼吸（breathe）。

我在长大
(I am growing bigger)

我吃（eat）得越多，长得越大。用不了多久，这层表皮（skin）就装不下我的身体了。这时，我会蜕掉旧皮，换上一层更大的表皮。每次变装我都会换个不同的（different）颜色。

7天　　12天　　18天

危险警报

有些种类的毛毛虫在察觉到危险来临时，就会迅速伸出橙色的触角，释放出臭味，把敌人吓跑。

有时，毛毛虫会把旧皮当成一顿大餐吃掉。

花海为家

蝴蝶几乎会出现在任何鲜花盛开的地方。哪里有花,哪里就有蝴蝶的家。春天是寻找蝴蝶卵的最佳季节,如果你仔细观察,就能找到它们。

蛋壳是我的第一顿大餐!

是时候孵化啦
(Time to hatch)

在卵内长了约5天后,我马上要孵化成一条小毛毛虫(caterpillar)了。我得不停地嚼啊嚼(chew),才能嚼出一条通往卵外的路,这可真累啊。

这个只有两天大的卵很快就会开始变色(change colour)。

毛毛虫宝宝要咬好几个小时(many hours)才能成功从卵壳里钻出来。

产卵

雌蝴蝶弯下（curve）柔软的（bendy）身体，在一片叶子上产卵。蝴蝶卵黏黏的，能粘在叶子上，不会滚落（roll away）。

甜蜜的家

每种蝴蝶只会在特定的几种植物上产卵，本书中的蝴蝶最喜欢在野生胡萝卜和大茴香的枝叶上产卵。

大茴香

野生胡萝卜

我出生之前
(Before I was born)

爸爸妈妈在田野（field）里飞舞时相遇了。它们彼此绕着对方飞了几圈，然后落在一朵花（flower）上交配。

交配完成后，雄蝴蝶飞走了，雌蝴蝶则会寻找一株植物（plant），并在它的叶子上产卵。

这是我的鳞片的特写。

蝴蝶的翅膀由成千上万片细小的鳞片（scale）组成。

翻开下一页，看看我是怎样长大的吧。

我是一只蝴蝶
(I'm a butterfly)

我扇动着五彩缤纷的翅膀（wing）在花丛中飞舞，用长而卷曲的"舌头"（tongue）吸食着花蜜（nectar）。

蝴蝶体表长有数百万根细小的绒毛（soft hair）。

蝴蝶的触角（antennae）能帮助蝴蝶分辨气味和保持平衡。

蝴蝶靠中空的细长口器（proboscis）吸食（suck）花蜜，口器伸展后就像一根吸管（straw）！

目录 Contents

和我们一起飞舞吧！

4~5
我是一只蝴蝶

6~7
我出生之前

8~9
是时候孵化啦

10~11
我在长大

12~13
我很饿

14~15
抓紧了

16~17
是时候破蛹而出啦

18~19
准备振翅而飞吧

20~21
生命循环，周而复始

22~23
我世界各地的朋友

24
词汇表

Original Title: Butterfly
Copyright © Dorling Kindersley Limited, 2003
A Penguin Random House Company

本书简体中文版授权由人民邮电出版社独家出版，仅限于中国境内（不包括香港、澳门、台湾地区）销售。未经出版者书面许可，不得以任何方式复制或发行本书中的任何部分。

伴我们成长，

For the curious
www.dk.com

DK 动物成长奥秘
看！我在长大（中英双语版）

蝴蝶

英国 DK 公司 ◎ 编
Tigger 赵泽源 毛思齐 岳艺 ◎ 译
鹰之舞 沈成 ◎ 审

人民邮电出版社
北京

DK 动物成长奥秘
看！我在长大（中英双语版）

小·猫

英国 DK 公司 ◎编
黄晨熙 胡悦熙 徐若曦 刘安迪 ◎译
鹰之舞 沈成 ◎审

人民邮电出版社
北京

目录 Contents

Original Title: Kitten
Copyright © Dorling Kindersley Limited, 2005
A Penguin Random House Company

本书简体中文版授权由人民邮电出版社独家出版,仅限于中国境内(不包括香港、澳门、台湾地区)销售。未经出版者书面许可,不得以任何方式复制或发行本书中的任何部分。

4~5
我是一只猫

6~7
我的爸爸和妈妈

8~9
我出生了

10~11
妈妈给我们洗澡

For the curious
www.dk.com

12~13
我两周大了

14~15
我去玩啦

16~17
打闹游戏

18~19
我们去探险吧

20~21
生命循环，周而复始

22~23
我世界各地的朋友

24
词汇表

我是一只猫
（I'm a cat）

我的生存能力很强。我长着柔软的毛发，这些毛发不仅能让我保持体温，还能保护我的身体。我喜欢蹦蹦跳跳（scamper）。我还会捉老鼠（mouse）呢！

又长又弯的尾巴（tail）能帮助我保持身体的平衡。

我的爪子很锋利，能帮助我捕食（hunt）和抓取。

这是我的家人和我的朋友们。

我即使在黑暗中（in the dark）也可以看得很清楚。

我的胡须（whiskers）也能帮助我保持身体平衡。

贪睡的小猫

我一天中的大多数时间都在睡觉，并且随时（any time）随地（any position）都能睡着。

翻开下一页，看看我们是怎样长大的吧！

我的爸爸和妈妈
（My dad and mum）

我的爸爸和妈妈曾经是邻居，它们第一次见面是在花园（garden）里。爸爸赶走了其他雄性（males）猫，之后它才和妈妈在一起并有了小宝宝。

保持卫生

我们喜欢舔自己的皮毛，保持身体干净和整洁。爸爸喜欢先舔（lick）爪子，再用爪子摩擦（rub）脸部。

这只猫妈妈的肚子（tummy）已经非常大了，因为猫宝宝快出生了。

这是我的妈妈。 这是我的爸爸。

我出生了
（I'm one day old）

刚出生的我非常饿（hungry），想喝奶了。猫妈妈用爪子轻轻推我们，帮助我们喝到它的乳汁，虽然有点挤（squeeze），但幸好我们都能适应（fit in）。

在小猫喝奶的时候，猫妈妈就安静地躺着（lie quietly）等着它们吃完。

刚出生的猫宝宝

猫宝宝刚出生的那几天里（the first few days），它们大部分时间都是叠在一起睡觉、取暖。

猫的小知识

- 🐈 新生的猫宝宝几乎什么也看不见,但它们能通过气味找到妈妈。

- 🐈 新生的猫宝宝几乎整天都在睡觉。

- 🐈 进食时,猫宝宝喉咙里会发出"呼噜呼噜"的声音,表示它们很开心。

妈妈给我们洗澡
（Mum washes us）

我们太小了，还不会自己清理身体，所以妈妈就舔遍我们的全身，帮助我们清洗（wash）身体。我们变得干干净净（clean），也感觉（feel）非常安心和幸福。

免费乘车

猫宝宝太小还不会走路，所以猫妈妈把它们轻轻地叼在嘴里，方便移动（move）它们。小猫会蜷起（tuck up）小短腿，乖乖地让妈妈叼着。

轮到我的姐妹洗澡了。

我两周大了
（I'm two weeks old）

我终于可以睁开（open）眼睛，看到美丽的世界了。现在，我开始蹒跚（wobbly）学步，而且听力也比刚出生时更敏锐了。

怎样了，现在我能看见我的兄弟姐妹了！

夜视

成年后的我们,夜视能力比人类更优秀。凭借着超级视力(super vision),我们能在夜间轻松捕食。

猫宝宝一直喵喵地叫,这样猫妈妈就能知道宝宝在哪里了。

瞧,这只猫宝宝正在练习怎样站起来(stand up)呢!

我去玩啦
（Now I can play）

我出生4周了，充满活力，喜欢和兄弟姐妹一起玩耍。我们用妈妈的尾巴来练习猛扑（pounce），但是我觉得妈妈并不喜欢我们这样做。

准备，准备，冲啊！

猫宝宝会对任何会动的东西感兴趣，并以此来练习捕食（hunt）。

妈妈的尾巴对我们来说是一个方便的玩具（handy toy）。

吃午饭

猫宝宝还需要母乳,但也已经可以吃固体食物(solid food)了。

我想,是时候带你们去外面玩耍了。

打闹游戏
（Play fighting）

我和我的兄弟在嬉戏打闹（fight）。但这只是一场游戏，不会受伤的。

我藏在草丛里，伺机而出。

扑来了！

当猫宝宝跳跃时，尾巴能够帮助它保持身体平衡（balance）。

两只猫宝宝只是在玩耍，所以不会使用锋利的爪子（claw）。

惊恐的猫宝宝

猫宝宝以前从来没有见过青蛙（frog），它被吓坏了（scared），于是它弓起背、竖起毛，让自己看上去更强大。

我们去探险吧
（Let's explore）

我们已经10周大了，对这个世界充满了好奇。这根树枝（branch）是个有趣的（fun）地方。爬上去（climbing up）简单，但是要下来就难多了。

从这儿可以下去吗？

快看我！我已经是12周大（12 weeks old）的大猫了。

四脚朝天

当猫四脚朝天从高处跌落时，会在空中快速翻转（turn around），四脚着地，安全落下。

生命循环，周而复始
The circle of life goes round and round

现在你知道我怎样长成一只成年猫了吧！

Now you know how I turned into a grown-up cat.

再见啦！我们要自个儿亲热去啦——互相抱抱，彼此呜呜。

我世界各地的朋友
My friends from around the world

暹罗猫（Siamese Cat）喜欢聊天，所以它们整天都在"喵喵"叫（meow）。

红波斯猫（Red Persian Cat）的毛发呈铁锈红色（rusty colour），它的毛长长的，如丝般顺滑（silky）。

你来说说看，为什么我

鲭鱼虎斑猫（Mackerel Tabby Oriental，虎斑猫的一种）喜欢后腿着地蹲坐在一起，相互依偎（cuddle）。

我的朋友们来自世界各地,它们体形（size）不同,毛色（colour）各异。

马恩岛猫（Manx Cat）的毛色丰富,但它们没有尾巴。

我是异国短毛猫。

叫作蓝色短毛猫呢？

猫的小知识

- 猫是用位于上颚的一个特殊的器官来闻气味的。
- 猫的行进方式比较特别,它是单侧前进的——两条左腿先前行,再换成两条右腿,左右两侧交替前行。
- 猫的一生中有三分之一的时间都在梳理毛发。

词汇表 Glossary

皮毛
Fur
覆盖在猫的体表的、能起保护作用的厚毛。

猛扑
Pounce
突然跃起去扑击某个物体或其他动物的行为。

一窝幼崽
Litter
同时出生的一窝猫宝宝。

喉音
Purr
猫在吃饱后或其他心情愉悦的时候所发出的"呼噜呼噜"声。

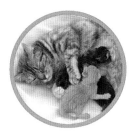

喂奶
Nurse
猫宝宝吮吸乳头，并从猫妈妈的乳房获取乳汁。

舌头
Tongue
猫的舌头表面粗糙不平可以用来直接舔食物或自己。

致谢 Acknowledgements

感谢以下人员及机构提供图片：
(Key: a=above; c=centre; b=below; l=left; r=right; t=top)
1 Powerstock: Frank Lukasseck c. 2-3 Zefa Visual Media: Frank Lukasseck. 3 Powerstock: Martin Rugner c. 4 Getty Images: Taxi cfl. 4-5 Steven Moore Photography. 4-5 Warren Photographic: Jane Burton x2. 5 Corbis: Helen King cl. 5 Getty Images: Stone tc; Taxi cra. 6 DK Images: Dave King cbl. 6 Steven Moore Photography: cl. 6-7 Steven Moore Photography. 7 DK Images: Jane Burton l, r. 8 DK Images: Jane Burton, bcl. 8-9 DK Images: Jane Burton. 8-9 Steven Moore Photography. 10 OSF/photolibrary.com: IPS Photo Index cb. 10 Warren Photographic: Jane Burton bcr. 11 Warren Photographic: Jane Burton c. 12 Warren Photographic: Jane Burton, cbl. 12-13 Steven Moore Photography. 12-13 Warren Photographic: Jane Burton. 13 Nature Picture Library Ltd: Pete Oxford tl. 14 Warren Photographic: Jane Burton cbr. 14-15 Steven Moore Photography. 14-15 Warren Photographic: Jane Burton. 15 Steven Moore Photography: tl. 15 Warren Photographic: Jane Burton tl. 16 Powerstock: age fotostock cla. 16-17 Powerstock: Martin Rugner/age fotostock. 17 DK Images: Geoff Brightling bc. 17 Steven Moore Photography: bcr. 17 Warren Photographic: Jane Burton br. 18-19 Warren Photographic: Jane Burton. 19 N.H.P.A.: Agence Nature r. 19 Warren Photographic: Jane Burton bl. 20 Alamy Images: Isobel Flynn c. 20 Warren Photographic: Jane Burton cla, ca, cra, crb, bc, bcl, bcr, car, cfl, cfr. 21 OSF/photolibrary.com: Dave Kingdon/IndexStock t. 21 Warren Photographic: Jane Burton cb. 22 DK Images: Dave King cla, cra; 22 Warren Photographic: Jane Burton tl. 22-23 DK Images: Marc Henrie. 23 DK Images: Dave King tr; Marc Henrie ca. 23 Powerstock: Martin Rugner/age fotostock br. 24 DK Images: Jane Burton cla, clb. 24 OSF/photolibrary.com: cbr. 24 Powerstock: Martin Rugner/age fotostock car. 24 Warren Photographic: Jane Burton cla, car.

其他图片版权属于多林·金德斯利公司。欲了解更多信息请访问DK Images网站。

Glossary

Fur
The thick coat of hair that covers and protects a cat.

Pounce
To leap or jump on an object or another animal, like a mouse.

Litter
A group of kittens that are all born at the same time.

Purr
The noise a cat makes when it is happy and full of food.

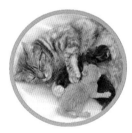
Nurse
When kittens drink milk from their mother's breast.

Tongue
Cats' tongues are very rough, so they can use them to lick.

Acknowledgements

感谢以下人员及机构提供图片：

(Key: a=above; c=centre; b=below; l=left; r=right; t=top)
1 Powerstock: Frank Lukasseck c. 2-3 Zefa Visual Media: Frank Lukasseck. 3 Powerstock: Martin Rugner c. 4 Getty Images: Taxi cfl. 4-5 Steven Moore Photography. 4-5 Warren Photographic: Jane Burton x2. 5 Corbis: Helen King cl. 5 Getty Images: Stone tc; Taxi cra. 6 DK Images: Dave King cbl. 6 Steven Moore Photography: cl. 6-7 Steven Moore Photography. 7 DK Images: Jane Burton l, r. 8 DK Images: Jane Burton, bcl. 8-9 DK Images: Jane Burton. 8-9 Steven Moore Photography. 10 OSF/photolibrary网站: IPS Photo Index cb. 10 Warren Photographic: Jane Burton bcr. 11 Warren Photographic: Jane Burton c. 12 Warren Photographic: Jane Burton, cbl. 12-13 Steven Moore Photography. 12-13 Warren Photographic: Jane Burton. 13 Nature Picture Library Ltd: Pete Oxford tl. 14 Warren Photographic: Jane Burton cbr. 14-15 Steven Moore Photography. 14-15 Warren Photographic: Jane Burton. 15 Steven Moore Photography: tl. 15 Warren Photographic: Jane Burton tl. 16 Powerstock: age fotostock cla. 16-17 Powerstock: Martin Rugner/age fotostock. 17 DK Images: Geoff Brightling bc. 17 Steven Moore Photography: bcr. 17 Warren Photographic: Jane Burton br. 18-19 Warren Photographic: Jane Burton. 19 N.H.P.A.: Agence Nature r. 19 Warren Photographic: Jane Burton bl. 20 Alamy Images: Isobel Flynn c. 20 Warren Photographic: Jane Burton cla, ca, cra, crb, bc, bcl, bcr, car, cfl, cfr. 21 OSF/photolibrary网站: Dave Kingdon/IndexStock t. 21 Warren Photographic: Jane Burton cb. 22 DK Images: Dave King cla, cra; 22 Warren Photographic: Jane Burton tl. 22-23 DK Images: Marc Henrie. 23 DK Images: Dave King tr; Marc Henrie ca. 23 Powerstock: Martin Rugner/age fotostock br. 24 DK Images: Jane Burton cla, clb. 24 OSF/photolibrary网站: cbr. 24 Powerstock: Martin Rugner/age fotostock car. 24 Warren Photographic: Jane Burton cla, car.

其他图片版权属于多林·金德斯利公司。欲了解更多信息请访问DK Images网站。

My cat friends from around the world come in lots of different colours and sizes.

Manx Cats come in many colours, but do not have a tail.

I'm a Dilute Calico Cat.

a Blue Shorthair Cat?

Kitten facts

- Cats smell with a special organ on the roof of their mouth.
- Cats step with both left legs, then both right legs when they walk or run.
- Cats spend one-third of their time grooming their fur.

My friends from around the world

Siamese Cats love to talk and will meow all day long.

The Red Persian Cat has long, silky hair in a solid rusty colour.

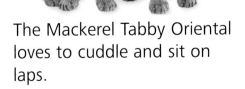

The Mackerel Tabby Oriental loves to cuddle and sit on laps.

Can you see why I am called

Bye bye, we're off for cuddles and purrs!

The circle of life goes round and round

Now you know how I turned into a grown-up cat.

Look at me now. I'm 12 weeks old and very grown up.

Head over heels

Cats almost always land on their feet. This is because they can turn around in the air.

Let's explore

We are ten weeks old and are very curious. This branch is a fun place to explore. Climbing up is easy, but it will be harder to climb down.

Is this the way down?

Here I come!

The kitten's tail helps her to balance while she leaps.

This is just play, so the kittens don't use their claws.

Scaredy cat

This kitten has never seen a frog before. He is scared, so he is puffing himself up and trying to look bigger.

Play fighting

My brother and I pretend to fight with each other. It's just a game, and no one gets hurt.

I hide in the grass and wait to pounce.

Lunch bunch
The kittens are still nursing, but they are also starting to eat solid food.

I think it's time for you to play outside...

Now I can play

I'm four weeks old and full of energy. My brothers and sisters and I love to play. We learn to pounce by practicing on mum's tail, but I don't think she likes it very much.

Ready, steady, go!
Kittens will play with almost anything that moves. This is how they learn to hunt.

Mum's tail makes a handy toy for the kittens.

Night sight

When they are grown up these kittens will see better at night than we can.
They will use this super vision to hunt at night.

This little kitten is learning to stand up.

The kittens call to mum so she always knows where they are.

I'm two weeks old

At last! My eyes are open so I can finally see where I am going. Now I can start to practice wobbly walking. I can also hear much better than I could when I was born.

Great! Now I can see my brothers and sisters!

It's my sister's turn to be cleaned.

Mum washes us

We are too little to wash ourselves, so mum does it by licking us all over. This not only keeps our fur clean, it also makes us feel safe and happy.

A free ride
Baby kittens can't walk, so when mum wants to move them, she picks them up gently in her mouth. They hold still and tuck up their legs.

Kitten facts

- The kittens can't see, but they can smell where mum is.
- Newborn kittens sleep most of the time.
- During feeding, the kittens will purr if they are getting enough milk.

I'm one day old

I'm hungry. It's time for me to nurse. Mum helps us all to find her milk by gently pushing us with her paws. It's a big squeeze but we all fit in.

The mother cat lies quietly while her litter drinks.

Newborns
For the first few days, the kittens spend almost all of their time lying together in a pile, keeping warm.

This is my mum... and this is my dad!

My dad and mum

My dad and mum live next door to each other. They met in the garden. Dad had to chase away a lot of other males before he had kittens with mum.

Keeping clean
Cats love to keep their fur clean and neat. Dad washes himself by licking his paws and then rubbing his face.

This mother cat's tummy is very big because it is almost time for her litter of kittens to be born.

Cats' eyes can see well in the dark.

Whiskers also help cats to balance.

Lullaby cat
Cats spend most of the day sleeping. They can sleep just about any time, and in any position.

Now turn the page and watch us grow...

I'm a cat

I can live almost anywhere. I have soft fur to keep me warm and protect me. I love to scamper, play, and hunt for mice.

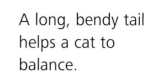

A long, bendy tail helps a cat to balance.

Sharp claws help cats to hunt and to grab things.

Here are mum and dad with all our friends and family.

12~13
I'm two weeks old

14~15
Now I can play

16~17
Play fighting

18~19
Let's explore

20~21
The circle of life

22~23
My friends from around the world

24
Glossary

Contents

4~5
I'm a cat

6~7
My dad and mum

8~9
I'm one day old

10~11
Mum washes us

DK 动物成长奥秘
看！我在长大（中英双语版）

青蛙

英国 DK 公司◎编

陈彦琦 潘弘毅 汪思彤 殷稼宸◎译

鹰之舞 沈成◎审

人民邮电出版社

北京

Original Title: Frog
Copyright © Dorling Kindersley Limited, 2003
A Penguin Random House Company

本书简体中文版授权由人民邮电出版社独家出版，仅限于中国境内（不包括香港、澳门、台湾地区）销售。未经出版者书面许可，不得以任何方式复制或发行本书中的任何部分。

随我跳跃，看我成长。

For the curious
www.dk.com

目录 Contents

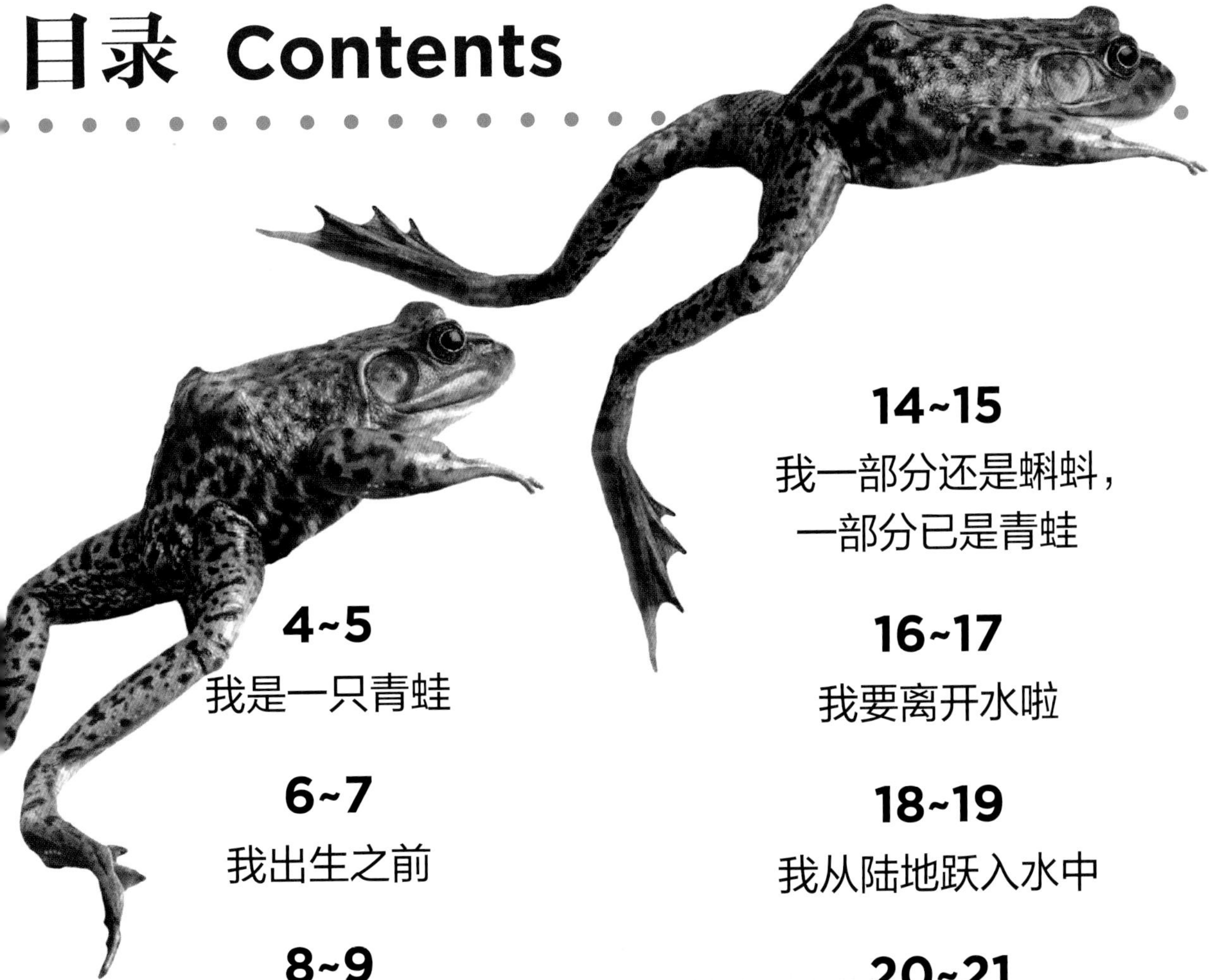

4~5
我是一只青蛙

6~7
我出生之前

8~9
我在卵中成长

10~11
我破卵而出啦

12~13
4周后,我长牙啦

14~15
我一部分还是蝌蚪,一部分已是青蛙

16~17
我要离开水啦

18~19
我从陆地跃入水中

20~21
生命循环,周而复始

22~23
我世界各地的朋友

24
词汇表

我是一只青蛙
（I'm a frog）

我喜欢"呱呱"叫。我是两栖动物，我既可以在水（water）中生活（live），又可以在陆地（land）上栖息。我有一对大长腿，跳高（leap）跳远全靠它。

我的皮肤柔软又湿滑，还可以帮助我呼吸。

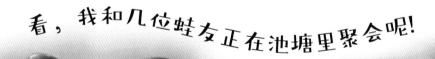

看，我和几位蛙友正在池塘里聚会呢！

大大的眼睛（eye）能帮助我更好地锁定快速移动的猎物。

我长着一对扁平的耳朵（ear）。

带蹼的脚（feet）让我能在水中快速游动。

哗啦！哗啦！

我喜欢生活在有水（water）的地方。池塘、小河或者泥泞的沼泽都是我的栖息之所。

翻开下一页，看看我是怎样长大的吧！

我出生之前
（Before I was born）

我的妈妈可厉害了！妈妈能在池塘（pond）里产下几千颗卵（egg），紧接着我的爸爸给卵受精（fertilized）。所以，在妈妈产卵时，爸爸必须离它很近。

结成一团的蛙卵被称为蛙卵块（frog spawn）。卵多力量大，想要把它们全部吃掉，可没那么容易！

呼叫青蛙同伴

蛙鸣（croaking）是青蛙间独有的交流方式。雄蛙（male frog）通过蛙鸣来吸引雌蛙（female frog）的注意。雌蛙也会根据雄蛙的鸣叫来分辨它们的位置。

蛙妈妈产卵时，蛙爸爸会紧紧地抱住它。

蛙妈妈的肚子（tummy）里满满都是卵。

蛙卵小知识

- 蛙妈妈一次产卵可多达4000颗。
- 蛙卵块中含有蝌蚪宝宝发育所需的营养。

我在卵中成长
（Now I'm growing inside my egg）

我在卵中努力地成长。渐渐地，我长出了尾巴（tail）和鳃（gill）。再过几天，我就能破卵而出了！

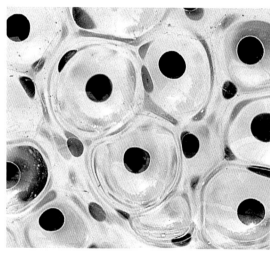

"蛙"之初

在生命（life）之初，我只是一枚米粒大的小黑卵（a tiny black egg），周围包裹着透明的胶质膜，使我免受意外碰撞（bump）造成的伤害！

9天后，蝌蚪马上就要孵化（hatch）出来了！

蛙卵小知识

- 蛙卵块会浮上水面,沐浴温暖的阳光。

- 胶质膜遇水膨胀,可以更好地保护卵。

- 大约10天后,蝌蚪就会被孵化出来。

10天后,我开始努力从周围黏糊糊的胶质中挣脱出来。

我破卵而出啦
（I'm ready to hatch out）

我破卵而出，奋力游向水面。那里不仅温暖（warm），而且还有充足的食物（food）。

数百只蝌蚪在同一时刻（at the same time）被孵化。

小蝌蚪们往往需要努力上一整天（all day），才能冲破蛙卵，获得自由。

孵化后，蝌蚪要休息几分钟（a few minutes）才会游走。

有鳃真好

刚出生的蝌蚪（tadpole）带有外鳃，和鱼一样只能用鳃呼吸（breathe）。鳃里遍布细小的血管，能从水中吸收氧气。

4周后，我长牙啦
(After four weeks my teeth begin to grow)

我终于可以吃昆虫（insect）了！当一只肥美的虫子落入水中时，我就能和兄弟姐妹一同美餐一顿啦。

蝌蚪们用小小的牙齿（teeth）咀嚼食物。

蝌蚪小知识

- 4周后,蝌蚪的外鳃被皮肤覆盖,成为内鳃,肺也开始发育。
- 大部分蝌蚪还没来得及长大就被捕食者吃掉了。

危机四伏

池塘里栖息的其他生物(creature),也在寻觅美食,因此我们必须游得足够快,才能从饥饿的鳟鱼(trout)口中逃脱!

对我来说,这些蝌蚪游得太快啦!

我一部分还是蝌蚪，一部分已是青蛙
（Now I'm part tadpole and part frog）

当我6周大（six weeks old）时，新长出的后肢让我在水中游得更加自如。现在的我一部分身体还是蝌蚪，一部分身体已经是青蛙了。我已经长成一只名副其实的幼蛙（froglet）了。

当蝌蚪开始用后肢代替尾巴划水（paddle）时，失去作用的尾巴就会逐渐变短。

瞧，我没有胳膊！

蝌蚪会先长出后肢（back leg），再长出前肢。同时，它的身体（body）和头部（head）也会变得越来越大。

这只小虫子真是太美味了!

幼蛙爱吃虫子(bug),比如这只松脆的划蝽(water boatman)。

幼蛙小知识

🐸 一只幼蛙长成青蛙大约需要14周。

🐸 幼蛙已经可以使用肺呼吸了,它们会迅速游到水面呼吸新鲜空气。

我要离开水啦
（Now I'm ready to leave the water）

3个月后（after three months），我已经能用肺（lung）和皮肤呼吸了。现在，我终于可以在陆地开启新生活啦！

尾巴，再见啦！

当我13周大（13 weeks old）时，我的尾巴基本消失了，只剩下一丁点儿。

我的个头非常小，小到可以坐在你的指尖上！

3年后，我的体形会和你的拳头一般大。

青蛙成长小知识

- 成蛙的体形比初次离开水面时增大了10倍。
- 夏季是观察新的成年青蛙的最佳时期。

我从陆地跃入水中
（I leap from land to water）

我需要时刻保持皮肤湿润（wet），才能依靠它辅助呼吸，维持健康（stay healthy）。因此，在岸上进食后，我还会再跳回水中游泳。

各就各位……预备……跳！

蛙跳小知识

- 如果你拥有青蛙般出色的跳跃能力，那么只需跳4次，你就能跳过整个足球场。

- 青蛙喜欢四处蹦跶。这样它们不仅能找到各种好吃的，还能躲避捕食者的袭击。

伸得长，够得远

青蛙的舌（tongue）根位置与人类的不一样，它位于蛙嘴（mouth）的前端，而不是后方。因此青蛙可以把舌头伸得很长去捕食美味的昆虫。

小心，我来啦……哗啦！

生命循环，周而复始
The circle of life goes round and round

现在你知道我怎样长成一只滑溜溜的青蛙了吧！

Now you know how I turned into a slippery frog.

呱呱！呱呱！我们明年春天见！

21

我世界各地的朋友
My friends from around the world

这只小小的丛蛙只比人类的指甲盖大一点。

绿油油的舌疣非洲树蛙（Tinker Reed Frog）是叫声最响亮的蛙之一。

来自澳大利亚的绿雨滨蛙（White's Tree Frog）至少可以存活21年。

俗称"箭毒蛙"（Poison Dart Frog）的丛蛙全身色彩鲜亮，可以警示敌人，让对方远离自己。

皮肤疙疙瘩瘩的角花蟾又

为了适应各自的生存环境，来自世界各地的青蛙朋友拥有不同的皮肤颜色（colour）和体形（shape）。

牛蛙（Bullfrog）的胃口非常大。

安东暴蛙俗称番茄蛙（Tomato Frog），它的身体像它的名字一样，又红又圆。

被称为"带腿的嘴"。

青蛙小知识

- 生活在非洲的巨谐蛙是世界上最大的青蛙，它可以长到30厘米长。
- 有些青蛙有毒，它们分泌的毒液足以让一个成年人丧命。
- 1977年，一只尖吻皱蛙创下了青蛙三级跳纪录——10米。

词汇表 Glossary

蹼
Webbing

连接青蛙脚趾之间的薄膜，能帮助青蛙更好地游泳。

鳃
Gills

蝌蚪用于呼吸的器官。

蛙卵块
Frog spawn

聚集在一起的很多蛙卵。

蝌蚪
Tadpole

青蛙刚从卵中孵化出来的幼体，有尾巴并且生活在水中。

孵化
Hatch

蝌蚪破卵而出的过程。

幼蛙
Froglet

介于蝌蚪和成蛙之间，四肢正在生长。

致谢 Acknowledgements

感谢以下人员及机构提供图片：

(Key: a=above; c=centre; b=below; l=left; r=right; t=top)
1: Getty images/David Aubrey c; 2-3: ImageState Pictor Ltd/Paul Wenham-Clarke; 4-5: Getty Images; 5: Getty Images tr; 6: Stuart R. Harrop bl; 7: FLPA - Images of Nature/Derek Middleton c; 8: N.H.P.A./Roger Tidman l; 9: N.H.P.A./Stephen Dalton c; 10-11: ImageState Pictor Ltd; 11: Stuart R. Harrop/Prof. br; 12: N.H.P.A./Stephen Dalton c; 13: Getty Images br; 14-15: N.H.P.A./G. I. Bernard; 16-17: Getty Images; 17: ImageState Pictor Ltd/Paul Wenham-Clarke r; 18-19: ImageState Pictor Ltd/Paul Wenham-Clarke; 19: N.H.P.A./Stephen Dalton cr; 20: N.H.P.A./Laurie Campbell c; Stephen Dalton tcl; 23: Jerry Young tr; 24: N.H.P.A./Roger Tidman cl.

Jacket Front: Jerry Young bc.

其他图片版权属于多林·金德斯利公司。欲了解更多信息请访问DK Images网站。

Glossary

Webbing
Thin skin that connects the frog's toes and helps it to swim.

Gills
A part of the tadpole's body that helps it to breathe.

Frog spawn
What frogs' eggs are called when they are all together.

Tadpole
A newly hatched frog. Tadpoles have tails, and live in the water.

Hatch
When a baby frog hatches, it comes out of its egg.

Froglet
In between a tadpole and a frog. Arms and legs are growing.

Acknowledgements

感谢以下人员及机构提供图片：

(Key: a=above; c=centre; b=below; l=left; r=right; t=top)
1: Getty images/David Aubrey c; 2-3: ImageState Pictor Ltd/Paul Wenham-Clarke; 4-5: Getty Images; 5: Getty Images tr; 6: Stuart R. Harrop bl; 7: FLPA - Images of Nature/Derek Middleton c; 8: N.H.P.A./Roger Tidman l; 9: N.H.P.A./Stephen Dalton c; 10-11: ImageState Pictor Ltd; 11: Stuart R. Harrop/Prof. br; 12: N.H.P.A./Stephen Dalton c; 13: Getty Images br; 14-15: N.H.P.A./G. I. Bernard; 16-17: Getty Images; 17: ImageState Pictor Ltd/Paul Wenham-Clarke r; 18-19: ImageState Pictor Ltd/Paul Wenham-Clarke; 19: N.H.P.A./Stephen Dalton cr; 20: N.H.P.A./Laurie Campbell c; Stephen Dalton tcl; 23: Jerry Young tr; 24: N.H.P.A./Roger Tidman cl.

Jacket Front: Jerry Young bc.

其他图片版权属于多林•金德斯利公司。欲了解更多信息请访问DK Images网站。

My frog friends all around the world come in different colours and shapes to help them survive in the place where they live.

The Bullfrog has a huge appetite.

This Tomato Frog looks like his name—red and round!

also called "a mouth with legs".

Fun frog facts

- The world's largest frog is the Goliath Frog of West Africa. It grows to 300mm (1 ft) long.

- Some frogs can kill a person with their poison.

- In 1977, a South African Sharp-Nosed Frog set a record for the froggie triple jump – 10 metres (33 feet 5 inches).

My friends from around the world

The tiny Tree Frog is as big as your fingernail.

The bright green Tinker Reed Frog is one of the loudest frogs.

The White's Tree Frog from Australia can live for more than 21 years.

This Poison Dart Frog is brightly coloured to warn enemies to stay away.

The Bumpy Horned Frog is

Croak, croak. See you next spring.

The circle of life goes round and round

Now you know how I turned into a slippery frog.

Long reach

A frog's tongue is attached to the front of its mouth, not the back, so it can reach farther for yummy bugs.

Look out, here I come... splash!

I leap from land to water

I need to keep my skin wet to stay healthy, so after eating on land, I leap back into the water for a swim.

Ready... Steady... GO!

Leaping facts

- If you could hop like a frog, you would be able to cross a football pitch in just four jumps.

- Frogs don't just hop for fun, they also hop to catch food and to escape from animals that want to eat them.

After three years I'll be as BIG as your fist.

Growing facts

- An adult frog is 10 times bigger than when it first came out of the water.

- Summer is the best time to spot new adult frogs.

Now I'm ready to leave the water

After three months, I breathe through my lungs and even my skin. Now I am ready to start living on land.

Going, going, gone!
When I am 13 weeks old my tail has almost gone. Just a tiny bit is left.

I'm so small, I can sit on your fingertip.

This bug will make a great snack.

Froglets eat bugs like this crunchy water boatman.

Froglet facts

- It takes about 14 weeks for a froglet to turn into a frog.

- The froglet now has lungs and breathes by quickly swimming to the surface to gulp some air.

Now I'm part tadpole and part frog

When I am six weeks old, my arms and legs start to grow. My new legs help me swim. I am part tadpole and part frog. I have become a froglet.

The tail gets shorter as the legs begin to paddle.

Look, no arms!
A froglet's back legs are the first to grow, and then its front legs grow. The froglet's body and head also start to grow much bigger.

Tadpole facts

- By four weeks, a tadpole's gills are covered in skin, and lungs begin to grow.
- Most tadpoles are eaten before they can grow up.

Danger lurks

Other creatures in the pond are also looking for food.
Tadpoles have to move fast to get away from a hungry trout.

These tadpoles are too quick for me!

After four weeks my teeth begin to grow

At last I can start eating insects. When a big, tasty worm drifts down from the surface I share it with my brothers and sisters.

Tadpoles use their tiny teeth to chew their food.

After hatching, tadpoles rest for a few minutes before swimming off.

Great gills
Tadpoles can breathe in the water because they have gills that can take in air that is in the water.

I'm ready to hatch out

Once I am out of my egg, I swim up to the surface of the pond where it's warm and there's plenty of food.

Hundreds of tadpoles hatch at the same time.

It takes all day for a tadpole to work its way out of the egg.

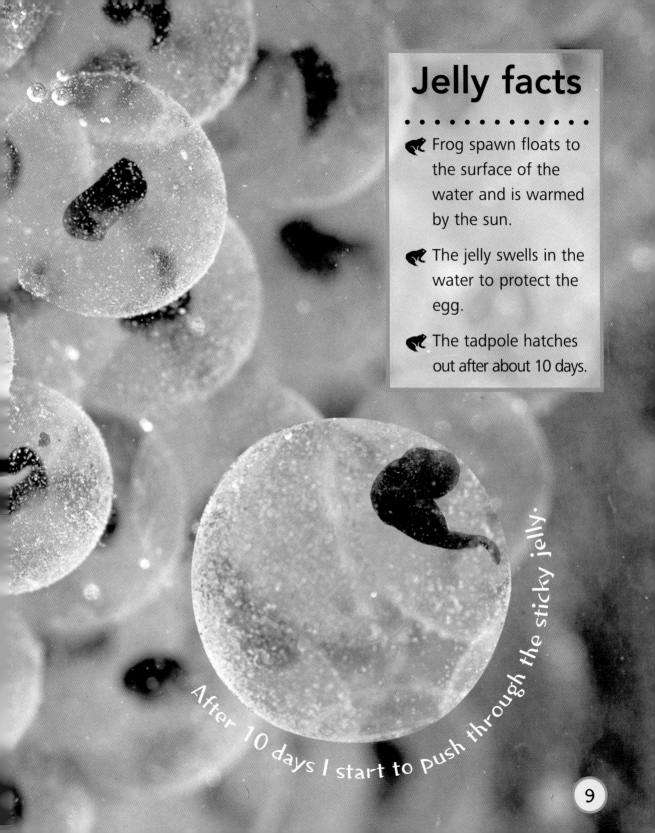

Jelly facts

- Frog spawn floats to the surface of the water and is warmed by the sun.
- The jelly swells in the water to protect the egg.
- The tadpole hatches out after about 10 days.

After 10 days I start to push through the sticky jelly.

Now I'm growing inside my egg

I already have my tail and gills. In a few days I will be big enough to wriggle out.

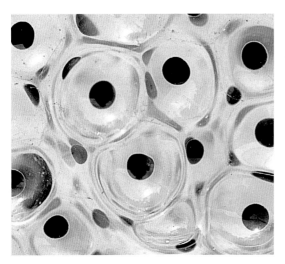

Beginning of life
A frog's life begins as a tiny black egg. The jelly around the egg protects it from bumps.

After nine days a tadpole is almost ready to hatch out.

Dad held onto mum while she laid her eggs.

The mum's tummy is full of eggs.

Spawn facts

🐸 The mother frog can lay as many as 4000 eggs.

🐸 Frog spawn have food inside them for the growing tadpole babies to eat.

Before I was born

Mum laid thousands of eggs in a pond, and then dad fertilized them. To do this, he had to stay very close to mum while she laid all her eggs.

A group of eggs is called frog spawn.

Calling all frogs
Croaking is the frog's version of talking. Croaking is how male frogs tell female frogs where they are.

DK WATCH ME GROW
FROG

DK 动物成长奥秘

看！我在长大（中英双语版）

农场动物

英国 DK 公司◎编

李如涵 郝悦言 黄致宁 李子祺◎译

鹰之舞 沈成◎审

人民邮电出版社

北京

Original Title: Farm Animals
Copyright © Dorling Kindersley Limited, 2005
A Penguin Random House Company

本书简体中文版授权由人民邮电出版社独家出版,仅限于中国境内(不包括香港、澳门、台湾地区)销售。未经出版者书面许可,不得以任何方式复制或发行本书中的任何部分。

目录 Contents

4~5
我是一只小鸡

6~7
长成一只公鸡

8~9
一只小羊羔

10~11
我2个月大了

For the curious
www.dk.com

12~13

刚出生的4头猪宝宝

14~15

泥水浴时间到了

16~17

棕色小牛

18~19

在绿油油的草地上

20~21

生命循环，周而复始

22~23

我世界各地的朋友

24

词汇表

我是一只小鸡
（I am a chick）

我是从妈妈产下的蛋里孵化出来的。我用尖尖的喙（beak）啄开了蛋壳。这可是很辛苦的活儿！

小鸡在被孵化出来之前，要在蛋里待上（stay）21天。

啄食

小鸡宝宝的喙非常坚硬，可以用来啄食草籽、谷物和虫子。

终于自由啦！

唧唧！唧唧！

挤啊，推啊！

干干的、毛茸茸的羽毛

鸡妈妈会一直给刚孵出来的小鸡保暖，直到（until）它的羽毛变得干干的（dry）、毛茸茸的（fluffy）。

长成一只公鸡
（Growing into a rooster）

我的羽毛开始产生变化（change）。2周大时，我开始长出成年鸡的羽毛和鸡冠（comb）。不久，我会成长为一只成年的公鸡。

8天大　　2周大　　4周大

换羽毛

小鸡成年后，羽毛会变得结实（strong）并能够防水（waterproof）。

喔喔喔!

8周大

公鸡

公鸡每天早上(morning)都会用响亮的(loud)打鸣声(crowing)唤醒整个农场。

一只小羊羔
（A little lamb）

我和我的孪生兄弟（twin brother）出生在春天。在我们身上的毛变干之前，妈妈会一直帮我们保持温暖和干燥。几分钟后，我们就准备站起来（stand up）了。

妈妈的帮助

羊妈妈先把小羊羔清理干净，然后用鼻子（nose）轻轻地推（nudge）它们，帮助小羊羔站起来。

我们在草堆上。

第一次喝奶

小羊羔是站着喝奶（nurse）的，在出生后的前4个月里，小羊羔一天要喝两次妈妈的乳汁（milk），直到它们能自己吃草为止。

多么暖和，多么舒适啊！

我2个月大了
（I'm two months old）

我长大了，可以到外面（outside）吃鲜嫩的青草了。我会待在妈妈身边，一旦遇到麻烦（in trouble），我就咩咩叫（bleat）。我的"羊毛外套"让我感到很暖和。

羊用上颚咀嚼（chew）食物。

羊蹄（hooves）上有两个脚趾。

羊毛外套

每年冬天,羊都会长出厚厚的羊毛。到了春天,这些羊毛将被剪掉并制成毛线。

我和我的家人、朋友生活在一个大牧场里。

在牧场

羊每天大部分时间待在露天的牧场里,寻找(looking for)青草作为食物。

刚出生的4头猪宝宝
（Four little piglets have just been born）

在舒适的猪棚（barn）里出生后不久，我们就可以开始进食和四处活动了，但我们必须待在妈妈附近。

新生的猪宝宝

猪妈妈将这头率先出生的猪宝宝舔（lick）干净。此时，猪宝宝已经睁开了眼睛（eye），几分钟后它就可以站起来了。

在这里，每头小猪都有足够的空间吃到妈妈的乳汁。

🐖 泥水浴时间到了
（Time to get mucky）

我们已经有4周大了，喜欢去户外活动。我们在草地上四处翻找（root around）食物。每日一次泥水浴可以帮助我们免受小虫子的骚扰。

需要帮助时，猪宝宝会高声尖声（squeal）呼唤妈妈。

猪用鼻子（snout）拱出植物的根（root），然后吃掉。

温馨的家

每个猪家庭(family)都有自己的小房子,叫作"方舟"(ark),可以挡风遮雨。到了晚上,小猪一家就会回到自己那舒适而又温暖的"方舟"里休息。

我喜欢哼叫着在泥巴里四处翻找好吃的。

棕色小牛
（The little brown calf）

我出生时，妈妈用粗糙的（rough）舌头把我舔干净。我饿了，但我需要站起来才能喝到奶。这时妈妈会轻推我一下，这样我就能站起来啦！

妈妈帮我站起来。

早餐时间

牛宝宝出生后的前6周只能喝妈妈的奶,所以它会一直紧跟着妈妈。

我有点儿站不稳……

牛宝宝正努力地站起来,可是它只能一条腿(leg)、一条腿地站起来。

我终于站起来啦!

站起来

你看!牛宝宝重心(weight)前倾,蹬直(straight)两条后腿。没过多久,它就靠自己站起来了。

在绿油油的草地上
（In the green grassy field）

我3个月大了，可以和牛群（herd）里的其他牛（cow）一起生活了。我们生活在一片满是青草和鲜花的田野上。

这是我和我最好的朋友。

牛都很友好（friendly），而且喜欢生活在一个大家庭里。

反刍

牛的胃分为4个部分。牛会把吃下的食物吐出来（spit up），进行二次咀嚼。这种将食物吐出来再次咀嚼的行为就叫作反刍（cud）。

牛可以吃很硬的草和植物，因为它们会通过反刍反复咀嚼这些食物。

我喜欢一整天都吃草，真香啊！

生命循环，周而复始
The circle of life goes round and round

……小鸡长成成年鸡

……牛犊长成牛……

现在你知道我们怎样从……

……小羊羔长成成年羊

……猪宝宝长成成年猪

咩……再见!

我世界各地的朋友
My friends from around the world

怀安多特鸡（Wyandotte Chicken）来自美国，产的蛋是棕色的。

格洛斯特郡花猪（Gloucester Old Spot Pig）喜欢吃苹果和橡子。

哎呦！哎呦！

阿尔萨斯猪（Alsace Pig）来自法国，它比一台冰箱还重。

曼克斯长角羊（Manx Longhorn Sheep）生活在马恩岛，它的头上长着长长的角。

我世界各地的农场朋友具有不同的个头（size）和体形（shape）。

温斯利代绵羊（Wensleydale Ram）长着长长的、蓬松的毛。

婆罗门牛（Brahmin Bull）生活在印度和巴西等炎热的地区。

我来自苏格兰。

唧唧唧唧

长毛可以帮助阿伯丁牛（Aberdeen Cow）御寒。

农场动物小知识

- 世界上鸡的数量比人类的数量还要多。
- 猪喜欢吃很多东西，比如草。
- 比起走下坡路，羊更喜欢走上坡路。
- 一头牛每年排泄的粪便能填满一栋房子。

词汇表 Glossary

喙
Beak

嘴周围又硬又尖的部分，鸟类用它来啄食。

口鼻部
Snout

某些动物长有被称为口鼻部的长长的鼻子，比如猪。

羽毛
Feather

鸟类身上的一层又轻又软的覆盖物，可以保暖。

乳房
Udder

母牛身上有乳头的部位，可以给小牛喂奶。

蛋
Egg

在被孵化（出生）之前，鸟类在蛋里生长。

蹄
Hoof

某些动物脚上长着的很硬的部分，比如猪或马。

致谢 Acknowledgements

感谢以下人员及机构提供图片：
Key: t = top, b = bottom, l = left, r = right, bkgrd = background, c = centre

Alamy/Juniors Bildarchiv: 2-3, /David Noton Photography: 11b, Archivberlin Fotoagentur GmbH/Bildagentur Geduldig: 14, 21 pig tl, /Minden Mas: 22-23b; Corbis/Najlah Feanny-Hicks: 10tr, /Ted Spiegel: 12-13, /Papilio/Steve Austin: 15t, /Robert Dowling 22tl, /David Katzenstein: 23tr; Country Life Picture Library/Joe Cornish: 16tl; Ecoscene/Angela Hampton: 8r, 9t, b, 21 lamb bl; Eye Ubiquitous/Hutchison: 8l; Getty Images/Image Bank/Cesar Lucas: 1, /Taxi/VCL: 4-5, /David Noton: 18l, 20 cow tr, /Stone/Tony Page: 7 bkgrd, /Peter Cade: 15b, 18-19c, /Photographer's Choice/Mike Hill: 7tr, /Lester Lefkowitz: 19r; Jeff Moore courtesy Wood & Sons of Hawkhurst: 10br.

其他图片版权属于多林·金德斯利公司。欲了解更多信息请访问DK Images网站。

Glossary

Beak
The hard, pointy part around the mouth that birds peck with.

Snout
The name for the long nose of some animals, including pigs.

Feather
A light, soft covering on a bird that keeps the bird warm.

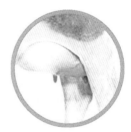
Udder
The part of the female cow that has the teats and gives milk.

Egg
Birds grow inside an egg until they hatch (are born).

Hoof
The hard part of the foot of some animals, such as pigs and horses.

Acknowledgements

感谢以下人员及机构提供图片：

Key: t = top, b = bottom, l = left, r = right, bkgrd = background, c = centre

Alamy/Juniors Bildarchiv: 2-3, /David Noton Photography: 11b, Archivberlin Fotoagentur GmbH/Bildagentur Geduldig: 14, 21 pig tl, /Minden Mas: 22-23b; Corbis/Najlah Feanny-Hicks: 10tr, /Ted Spiegel: 12-13, /Papilio/Steve Austin: 15t, /Robert Dowling 22tl, /David Katzenstein: 23tr; Country Life Picture Library/Joe Cornish: 16tl; Ecoscene/Angela Hampton: 8r, 9t, b, 21 lamb bl; Eye Ubiquitous/Hutchison: 8l; Getty Images/Image Bank/Cesar Lucas: 1, /Taxi/VCL: 4-5, /David Noton: 18l, 20 cow tr, /Stone/Tony Page: 7 bkgrd, /Peter Cade: 15b, 18-19c, /Photographer's Choice/Mike Hill: 7tr, /Lester Lefkowitz: 19r; Jeff Moore courtesy Wood & Sons of Hawkhurst: 10br.

其他图片版权属于多林·金德斯利公司。欲了解更多信息请访问DK Images网站。

Our farm friends from around the world come in lots of different sizes and shapes.

Wensleydale Rams have very long, shaggy wool.

The Brahmin Bull lives in hot places like India and Brazil.

I'm from Scotland.

cheep cheep

Aberdeen Cows have long hair to protect them from the cold.

Farm facts

- There are more chickens in the world than people.
- Pigs like to eat many things, such as weeds.
- Sheep prefer to walk uphill, rather than downhill.
- One cow makes enough manure each year to fill up your house.

My friends from around the world

The Wyandotte Chicken from America lays brown eggs.

Gloucester Old Spot Pigs live in England and love to eat apples and acorns.

Ouch! Ouch!

Alsace Pigs are from France and weigh more than a refrigerator!

Manx Longhorn Sheep live on the Isle of Man and have very long horns.

The circle of life goes round and round

... chick to chicken

Now you know how we grew from a ...

calf to cow

Chewing cud

Cows have a four-part stomach. They can spit up their food and chew it a second time. This food is called cud.

Cows can eat tough grass and plants because they chew their food over and over.

I love to chew grass all day long, yum yum…

In the green grassy field

I'm three months old and now I can spend my days with the other cows in our herd. We all live together in a big field full of grass and flowers.

This is me... and my best friend.

Cows are very friendly and like to live in a big herd.

Breakfast time

For the first six weeks, the calf will drink only his mother's milk. He stays close to mum at all times.

I'm a bit wobbly...

The calf gets up one leg at a time.

I'm up at last!

Getting up

The calf leans forward with all his weight, and pushes his back legs straight. In no time at all, he's up on his feet.

The little brown calf

When I am born, my mum cleans me with her rough tongue. I am hungry, but I need to stand up to feed. Mum gives me a helping nudge, and soon I can stand.

Mum gives me a boost.

Home sweet home

Each pig family has its own small house, called an ark. At night, the family goes into its ark, where it's cosy and warm.

I love to snuffle around in the mud for tasty roots.

Time to get mucky

We're four weeks old and we love to go outside. We root around in the grass, looking for food. A daily mud bath keeps bugs away.

The piglets will squeal if they need any help from Mum.

Pigs use their snouts to dig up roots to eat.

Four little piglets have just been born

We are born in a cosy barn. Right away, we are ready to start eating and moving around, but we make sure we stay close to mum.

Newborn piglet

This piglet was the first one to be born. Mum has licked him clean. His eyes are already open and in a few minutes he will be ready to stand up.

There is plenty of room for everyone

Woolly coat

The sheep grows a thick coat of wool every winter. In the spring, this wool is sheared off and made into yarn.

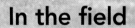

I live in a big field with my friends and family.

In the field

Sheep spend most of their day outside in the field, looking for tender grass to eat.

I'm two months old

I'm old enough to go outside and start eating tender grass. I stay close to mum and bleat if I am in trouble. My woolly coat keeps me warm.

Sheep chew with the roof of their mouth.

Sheep have hooves that are split into two toes.

First drink

Lambs nurse standing up. The lambs will drink their mother's milk for about four months. The lambs will nurse twice a day until they are old enough to eat grass.

and cosy in this straw.

A little lamb

My twin brother and I are born in the spring. Mum keeps us warm and dry until our wool dries. After a few minutes, we are ready to stand up.

With Mum's help
The mother sheep cleans off the lambs. Then she helps them to their feet by gently nudging them with her nose.

We are warm

Cock-a-doodle-doo

Eight weeks old

Rooster
The rooster wakes the farm up every morning with his loud crowing.

Growing into a rooster

My feathers are starting to change. When I am two weeks old, my adult feathers and my comb start to grow. Soon I will be a tall rooster.

Eight days old

Two weeks old

Four weeks old

Changing feathers
The chick's adult feathers are stronger and more waterproof.

Pecking for food
Chickens eat seeds, grain, and insects, which they peck at with their hard beaks.

Cheep cheep cheep

Free at last!

Squeeze and push

Dry and fluffy
The hen keeps the chick warm until its feathers are dry and fluffy.

I am a chick

I hatched out of an egg that my mother laid. I pecked my way out of the shell with my beak. It was very hard work.

The chick stays in its egg for 21 days before hatching.

12~13
Four little piglets have just been born

14~15
Time to get mucky

16~17
The little brown calf

18~19
In the green grassy field

20~21
The circle of life

22~23
My friends from around the world

24
Glossary

Contents

4~5
I am a chick

6~7
Growing into a rooster

8~9
A little lamb

10~11
I'm two months old

DK WATCH ME GROW
FARM ANIMALS

DK 动物成长奥秘
看!我在长大(中英双语版)

大象

英国 DK 公司◎编

张浩岩 曹添文 Natalie Wang 诸葛越◎译

鹰之舞 沈成◎审

人民邮电出版社

北京

Original Title: Elephant
Copyright © Dorling Kindersley Limited, 2005
A Penguin Random House Company

本书简体中文版授权由人民邮电出版社独家出版，仅限于中国境内（不包括香港、澳门、台湾地区）销售。未经出版者书面许可，不得以任何方式复制或发行本书中的任何部分。

For the curious
www.dk.com

目录 Contents

4~5
我是一头大象

6~7
我的爸爸和妈妈

8~9
我出生一个小时啦

10~11
泥浆浴

12~13
我3个月大了

14~15
过河

16~17
学会自己进食

18~19
我5岁了

20~21
生命循环,周而复始

22~23
我世界各地的朋友

24
词汇表

跟着我们一起走,见证我们的成长。

我是一头大象
（I'm an elephant）

我是一个大块头，我和一群大块头生活在一起。我能用长长的象鼻（trunk）卷起食物放进嘴里，还可以用鼻子取水。

我的皮肤布满皱褶，看起来很粗糙（rough）。

我的脚底长有厚厚的跟垫（padding）。

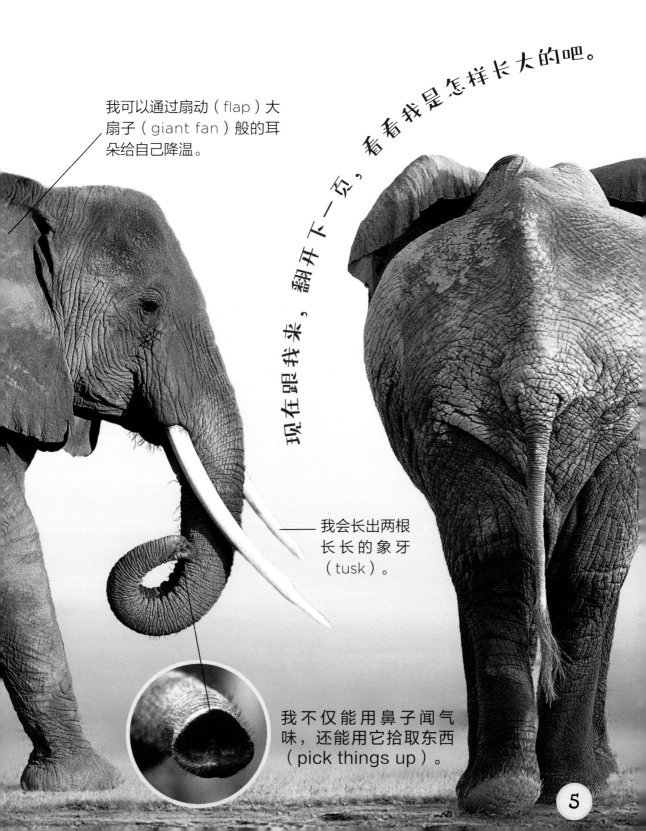

我可以通过扇动（flap）大扇子（giant fan）般的耳朵给自己降温。

现在跟我来，翻开下一页，看看我是怎样长大的吧。

我会长出两根长长的象牙（tusk）。

我不仅能用鼻子闻气味，还能用它拾取东西（pick things up）。

我的爸爸和妈妈
（My dad and mum）

我的妈妈和其他大象妈妈、大象宝宝一起生活在一个象群家族里（live in a herd），而我的爸爸则喜欢独自生活。我的爸爸和妈妈是在小水塘（water hole）边相识的。

这是我的妈妈。

象群的生活

象群通常由最年长的（the oldest）母象担任首领（leader）。它会带领象群寻找水源和食物。

这是我的爸爸——看它那对大耳朵！

成年公象喜欢独自（alone）生活，或者和一小群的成年公象一起活动。

我出生一个小时啦
（I'm one hour old）

我要在妈妈的肚子里孕育将近两年才能来到这个世界上。我刚出生时，其他大象都会在我的身边保护（protect）我，直到我可以站起来行走为止。

刚刚出生的我

我刚刚出生时身上裹有胎衣，有些脏乱（messy），妈妈会帮我在站起来前把胎衣清理干净。

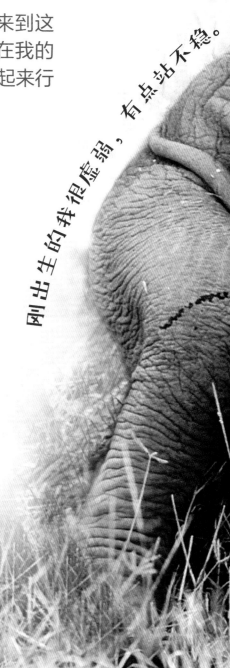

刚出生的我很虚弱，有点站不稳。

妈妈小心翼翼地帮（help）我站起来。

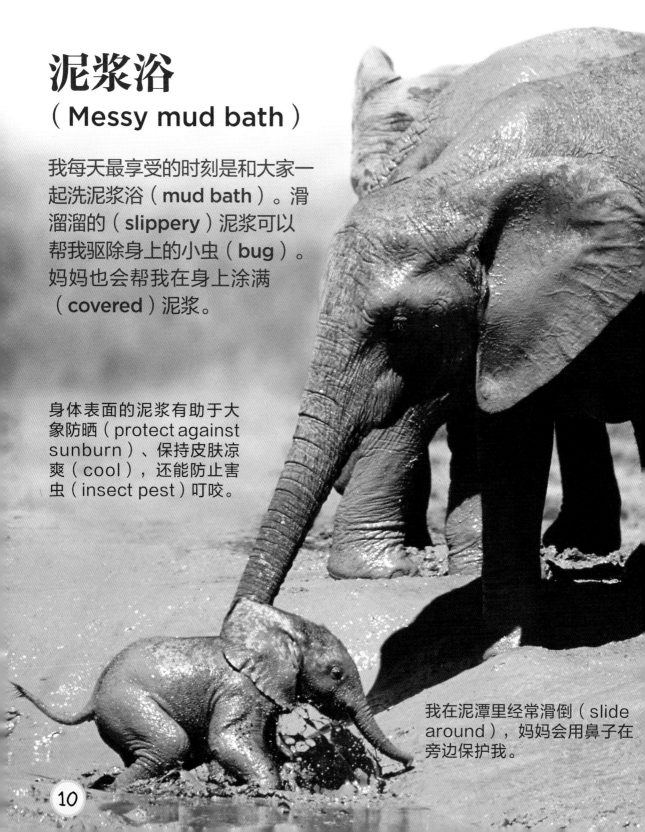

泥浆浴
（Messy mud bath）

我每天最享受的时刻是和大家一起洗泥浆浴（mud bath）。滑溜溜的（slippery）泥浆可以帮我驱除身上的小虫（bug）。妈妈也会帮我在身上涂满（covered）泥浆。

身体表面的泥浆有助于大象防晒（protect against sunburn）、保持皮肤凉爽（cool），还能防止害虫（insect pest）叮咬。

我在泥潭里经常滑倒（slide around），妈妈会用鼻子在旁边保护我。

上岸

泥潭的岸边有点陡峭（steep），我还要再长大一些才能自己爬上去，现在只能让妈妈推着我才能爬上岸。

我3个月大了
（I'm three months old）

我喜欢和其他小象一起玩耍和摔跤（wrestle），喜欢不停地甩动（flopping around）鼻子，我迫不及待地想变得更强壮有力，这样才能好好使用我的鼻子。

大象小知识

 成年大象通常站着睡觉，但大象宝宝有时候会躺着睡。

 一头成年大象每天排泄的粪便（poo）比一个小朋友的体重还重。

 大象宝宝大约在5岁前都会喝妈妈的奶。

危险!

成年大象的体形巨大,没有什么天敌(enemy),但是小象还是会有被鳄鱼、猎豹等动物攻击(attack)的危险。

鳄鱼

猎豹

过河
（Crossing the river）

我6个月大了，当雨季（rain season）来临，河水（river）开始上涨，我们有时需要过河才能吃到更可口的草（grass）。过河的时候，我会紧紧跟在（stay close to）妈妈身后。

大象的鼻子是一根天然的呼吸管。

在水深（deep）的地方，成年大象会用鼻子抬起（lift）小象，推着（push）它们过河。

水中嬉戏

大象虽然体形巨大,但是非常善于游泳,也很喜欢游泳,有时候甚至会到大海(ocean)里"一展拳脚"。

学会自己进食
（Learning to feed myself）

虽然我还在喝妈妈的奶，但是我已经学会用象鼻子捡东西吃了。我们一家每天要花好几个小时来进食。我们主要吃青草、树枝、种子（seed）和果实（fruit）。

小象只能吃靠近地面（ground）的食物。

成年大象可以从树上摘树叶吃。

卷一卷，拔一拔

大象可以用鼻子摘取树叶（pluck the leaves），也可以用它把美味的植物拔出来（pull up）吃掉。看，这头大象正想要用鼻子剥下枝条上的树皮（bark）来品尝。

我5岁了
（I'm five years old）

现在我有了一个妹妹。我会帮妈妈教它如何生活，就像小时候其他大象教我的那样。再过10年（ten years）我才能真正成年（grown up）。

这是我。

长大

大象的成长速度十分缓慢（slowly），寿命长达60~80年。公象会在13岁左右离开象群（leave the herd）。

我和妹妹一起跟着妈妈散步。

生命循环，周而复始
The circle of life goes round and round

现在你知道我怎样成长为一头成年大象了吧！

Now you know how I turned into a grown-up elephant.

再见，我们会很快再见的！

我世界各地的朋友
My friends from around the world

我的朋友有亚洲象和非洲象。

亚洲象（Asian Elephant）可以在丛林中帮助人类搬运沉重的木头（heavy log）。

亚洲象的耳朵较小，头部有隆起。

非洲象(African Elephant)长着长长的象牙和大大的耳朵。

来兜兜风吧!

我是一头非洲象宝宝。

这头亚洲象披上了节日的(festival)盛装。

大象小知识

- 大象每天大约花16个小时进食,却只用3~5个小时睡觉。

- 大象通过耳朵散热,耳朵中流动的血液(blood)会把体内的热量带走并散发出去。

- 大象与儒艮(dugong)有亲缘关系,儒艮是一种生活在浅海区域的海牛目动物(sea cow)。

词汇表 Glossary

公象
Bull
雄性大象的别称。

象群
Herd
一起生活和行动的大象家族。

母象
Cow
雌性大象的别称。

象皮
Hide
大象粗糙且凹凸不平的皮肤。

小象
Calf
不到5岁的幼象。

象牙
Tusks
大象突出嘴外的长长尖尖的牙齿，可以用于挖掘及获取食物。

致谢 Acknowledgements

感谢以下人员及机构提供图片：

(Key: a=above; c=centre; b=below; l=left; r=right; t=top)
1 DigitalVision: Gerry Ellis c. 2-3 Still Pictures: Dianne Blell.
4 DigitalVision: Gerry Ellis cfl. 4-5 DigitalVision: Gerry Ellis.
5 OSF/photolibrary.com: Gerry Ellis/Digital Vision fr. 5 Zefa Visual Media: Steve Craft/Masterfile bcl. 6-7 Getty Images: Taxi/Stan Osolinski. 7 Andy Rouse Wildlife Photography: c.
8 OSF/photolibrary.com: Martyn Colbeck cl. 8-9 OSF/photolibrary.com: Martyn Colbeck. 10-11 OSF/photolibrary.com: Peter Lillie.
11 ImageState/Pictor: bcr. 11 Science Photo Library: Tony Camacho tr. 12-13 DigitalVision: Gerry Ellis. 13 DK Images: Irv Beckman crb; Jerry Young cr.
14 FLPA: Frans Lanting/Minden Pictures car. 14-15 N.H.P.A.: Martin Harvey. 15 Getty Images: Cousteau Society/The Image Bank tcl. 16 Corbis: Jeff Vanuga bl. 16 DigitalVision: Gerry Ellis c. 16-17 FLPA: David Hosking.
17 N.H.P.A.: Ann & Steve Toon car. 18 Bruce Coleman Ltd: cl. 18-19 Alamy Images: Steve Bloom. 20 Alamy Images: Martin Harvey cl. 20 Corbis: Martin Harvey/Gallo Images cb. 20 DigitalVision: Gerry Ellis c. 20 DK Images: Shaen Adey cbl. 20 ImageState/Pictor: Nigel Dennis bcr. 20 OSF/photolibrary.com: IFA-Bilderteam Gmbh cra; Martyn Colbeck ca, car. 20 Andy Rouse Wildlife Photography: tl. 20 Safari Bill Wildlife Photography: crb. 21 DigitalVision: c. 22 Ardea.com: Tom & Pat Leeson bl. 22 DK Images: Dave King c. 22 DigitalVision: Gerry Ellis tl.
22-23 Corbis: Paul Almasy. 23 DigitalVision: Gerry Ellis tr, br.
23 OSF/photolibrary.com: Peter Lillie cr. 24 DigitalVision: car; Gerry Ellis cla, cr, cbr. 24 FLPA: cl. 24 OSF/photolibrary.com: Martyn Colbeck cbl.

其他图片版权属于多林·金德斯利公司。欲了解更多信息请访问DK Images网站。

Glossary

Bull
Another name for a male elephant who is fully grown.

Herd
A group of elephants who live and travel together.

Cow
A female who is fully grown and can have babies.

Hide
An elephant's skin, which is very rough and bumpy.

Calf
A very young elephant who is less than five years old.

Tusks
Long, sharp teeth used for digging and gathering food.

Acknowledgements

感谢以下人员及机构提供图片：

(Key: a=above; c=centre; b=below; l=left; r=right; t=top)
1 DigitalVision: Gerry Ellis c. 2-3 Still Pictures: Dianne Blell.
4 DigitalVision: Gerry Ellis cfl. 4-5 DigitalVision: Gerry Ellis.
5 OSF/photolibrary网站: Gerry Ellis/Digital Vision fr. 5 Zefa Visual Media: Steve Craft/Masterfile bcl. 6-7 Getty Images: Taxi/Stan Osolinski. 7 Andy Rouse Wildlife Photography: c.
8 OSF/photolibrary网站: Martyn Colbeck cl. 8-9 OSF/photolibrary网站: Martyn Colbeck. 10-11 OSF/photolibrary网站: Peter Lillie.
11 ImageState/Pictor: bcr. 11 Science Photo Library: Tony Camacho tr.
12-13 DigitalVision: Gerry Ellis. 13 DK Images: Irv Beckman crb; Jerry Young cr. 14 FLPA: Frans Lanting/Minden Pictures car. 14-15 N.H.P.A.: Martin Harvey. 15 Getty Images: Cousteau Society/The Image Bank tcl.

16 Corbis: Jeff Vanuga bl. 16 DigitalVision: Gerry Ellis c. 16-17 FLPA: David Hosking. 17 N.H.P.A.: Ann & Steve Toon car. 18 Bruce Coleman Ltd: cl. 18-19 Alamy Images: Steve Bloom. 20 Alamy Images: Martin Harvey cl. 20 Corbis: Martin Harvey/Gallo Images cb. 20 DigitalVision: Gerry Ellis c. 20 DK Images: Shaen Adey cbl. 20 ImageState/Pictor: Nigel Dennis bcr. 20 OSF/photolibrary网站: IFA-Bilderteam Gmbh cra; Martyn Colbeck ca, car. 20 Andy Rouse Wildlife Photography: tl. 20 Safari Bill Wildlife Photography: crb. 21 DigitalVision: c. 22 Ardea网站: Tom & Pat Leeson bl. 22 DK Images: Dave King c. 22 DigitalVision: Gerry Ellis tl. 22-23 Corbis: Paul Almasy. 23 DigitalVision: Gerry Ellis tr, br. 23 OSF/photolibrary网站: Peter Lillie cr. 24 DigitalVision: car; Gerry Ellis cla, cr, cbr. 24 FLPA: cl. 24 OSF/photolibrary网站: Martyn Colbeck cbl.

其他图片版权属于多林・金德斯利公司。欲了解更多信息请访问DK Images网站。

The African Elephant has long tusks and very big ears.

Come for a ride!

I'm a baby African Elephant.

This Asian Elephant has been decorated for a festival.

Elephant facts

- Elephants spend about 16 hours a day eating, and only 3 to 5 hours sleeping.

- Elephants cool off using their ears. Blood moves around their ears and cools off as it goes.

- The elephant is related to the dugong, a kind of sea cow that lives in shallow waters.

My friends from around the world

There are two types of elephant: Asian and African.

Asian Elephants are often used to lift heavy logs in the jungle.

Asian Elephants have small ears and bumpy heads.

The circle of life goes round and round

Now you know how I turned into a grown-up elephant.

My baby sister and I walk with mum.

I'm five years old

Now I have a baby sister. I will help teach her all about how to be an elephant, just like the other elephants taught me. It will be ten more years before I am all grown up.

This is me.

Growing up

Elephants grow very slowly and can live to be 60 or even 80 years old. Male elephants leave the herd when they are around 13.

The tall adults pluck leaves from the trees.

Twist and pull

Elephants use their trunks to pluck leaves and to pull up tasty plants. This elephant is going to chew the bark off this branch.

Learning to feed myself

I still drink milk from my mum, but now I can pick up food with my trunk too. My family spends hours eating every day. We eat grass, plants, seeds, and fruit.

The young elephants eat food that's found close to the ground.

Underwater fun

Elephants may be big, but this does not stop them from being strong swimmers. They enjoy swimming and will even swim in the ocean.

Crossing the river

When I'm six months old, the rains come and the rivers fill up. We have to cross to get to the better grass on the other side. I stay close to mum while we cross.

In deep water, adult elephants help the babies by lifting and pushing them through the water.

The trunk makes a handy snorkel

Danger!

Adult elephants are too big to have many enemies, but baby elephants are in danger of attacks.

Crocodile

Cheetah

I'm three months old

I like to play and wrestle with other young elephants. My trunk keeps flopping around, and I can't wait until I am strong enough to use it.

Elephant facts

 Adult elephants usually sleep standing up. Baby elephants sometimes lie down to sleep.

 Every day, an adult elephant can make a pile of poo that weighs more than you!

 The baby will drink his mother's milk for about five years.

Up we go...

Mum pushes her calf up the steep bank. It will be many weeks before he is strong enough to do this by himself.

Messy mud bath

My favourite time of day is when we all go for a mud bath. The mud is slippery, but it keeps the bugs away. Mum makes sure I am all covered in mud.

Mud protects against sunburn, and keeps elephants cool and free from insect pests.

The tiny calf slides around in the mud, but mum stays nearby to lend a steady trunk.

Mum gently helps her baby to his feet.

I'm one hour old

I grew inside my mother for nearly two years. After I'm born, other elephants stand nearby to protect me until I'm strong enough to walk.

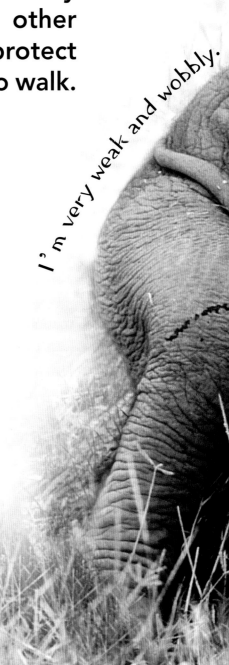

I'm very weak and wobbly.

Just born
Newborn elephants are very messy, so mum has to clean her baby off before he can stand up.

and this is my dad – look at his BIG ears!

Male elephants live alone or with small groups of other males.

My dad and mum

My mum lives in a herd with other mothers, babies, and young elephants. Dad likes to live on his own. Mum and dad met at the water hole.

This is my mum...

Living in a herd
The leader of the herd is the oldest female elephant. She leads the herd to where water and food can be found.

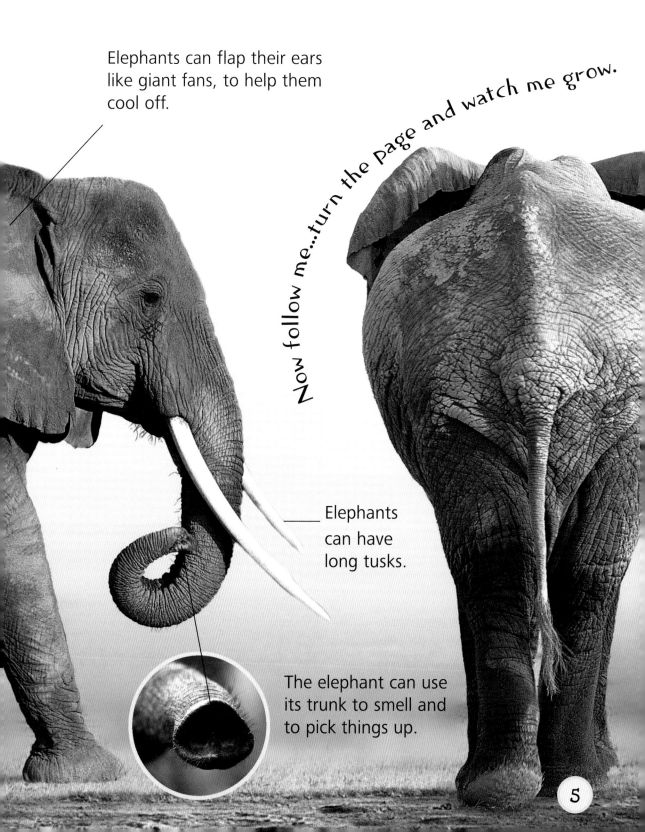

Elephants can flap their ears like giant fans, to help them cool off.

Now follow me...turn the page and watch me grow.

Elephants can have long tusks.

The elephant can use its trunk to smell and to pick things up.

I'm an elephant

I am very big. I live with a lot of other elephants in a herd. I use my long trunk to put food and water into my mouth.

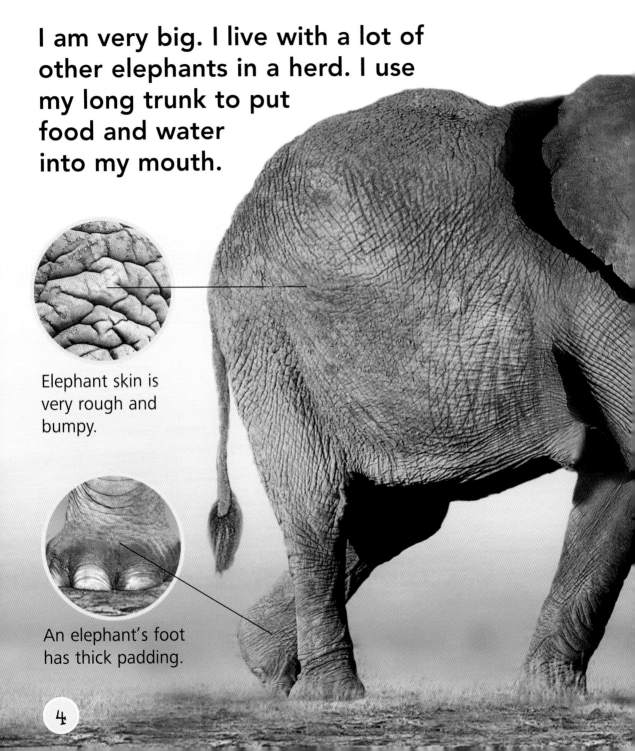

Elephant skin is very rough and bumpy.

An elephant's foot has thick padding.

14~15
Crossing the river

16~17
Learning to feed myself

18~19
I'm five years old

20~21
The circle of life

22~23
My friends from around the world

24
Glossary

Come walk with us and watch us grow!

Contents

4~5
I'm an elephant

6~7
My dad and mum

8~9
I'm one hour old

10~11
Messy mud bath

12~13
I'm three months old

DK WATCH ME GROW
ELEPHANT

DK 动物成长奥秘
看！我在长大（中英双语版）

大熊猫

英国 DK 公司 ◎编
吴含章 甘芸菲 谈知远 周正 ◎译
鹰之舞 沈成 ◎审

人民邮电出版社
北京

Original Title: Panda
Copyright © Dorling Kindersley Limited, 2008
A Penguin Random House Company

本书简体中文版授权由人民邮电出版社独家出版，仅限于中国境内（不包括香港、澳门、台湾地区）销售。未经出版者书面许可，不得以任何方式复制或发行本书中的任何部分。

目录 Contents

4~5
我是一只大熊猫

6~7
这是我的家

8~9
这是我的妈妈

10~11
我两个月大喽

FSC® C018179

For the curious
www.dk.com

12~13
勇攀新高

14~15
吃大餐啦

16~17
游戏时间

18~19
乐在成都

20~21
生生不息

22~23
生命循环，周而复始

24
词汇表

我是一只大熊猫
（I'm a giant panda）

我是熊（bear），不是猫（cat），我的家乡是中国。我有一身黑白相间的皮毛（fur）。我每天得花半天时间吃竹子（bamboo）。如果你要问我剩下的半天时间做什么。睡一睡（sleep），玩一玩（play），这就是我快乐的一天！

超级感官

大熊猫的嗅觉（smell）很灵敏，它们的听力（hearing）也超强，一听到有人靠近，它们立马开溜（leave），想发现它们可难着呢！

大熊猫的皮毛相当厚实（thick），并且带有油脂，所以它们的皮毛能够抵御寒冷（cold）又潮湿（wet）的天气。

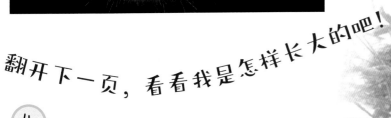

翻开下一页，看看我是怎样长大的吧！

毛茸茸的脚掌（paw）让大熊猫在滑溜溜的（slippery）雪地和陡峭的岩石（rock）上也能行走自如。

大熊猫的牙齿（teeth）可厉害了，坚硬的竹子都能咬穿呢！

这是我的家
（This is my home）

大熊猫只生活在中国，它们栖息在生长着丰富竹子的山地（mountains）。世界范围内，只有中国才有野生大熊猫。竹子是大熊猫最爱吃的食物（favourite food）。

不一样的家

晶晶（Jingjing）出生在中国成都大熊猫繁育研究基地（panda base）。那里也是它的爸爸、妈妈生活的地方。晶晶由专业的饲养员（keeper）照顾。

该休息了

大熊猫一般在竹林（bamboo grove）或者岩穴（rock den）里休息。年幼的大熊猫也会在树上休息（rest）。

山里的冬天很冷，常常是一片冰天雪地。

🐼 这是我的妈妈
（This is my mum）

晶晶的妈妈叫娅娅（Yaya）。晶晶一出生就被饲养员抱去称重（weight）和体检（check）了，但很快又被送回了娅娅身边，开始享受妈妈的照料了。

熊猫小知识

🐼 熊猫妈妈通常每次能生1~2只熊猫宝宝。

🐼 在出生后一年半（甚至更久）的时间里，熊猫宝宝都会跟妈妈待在一起。

🐼 熊猫宝宝成长为成年大熊猫大约需要5年时间。

妈妈抱一抱，烦恼全忘掉！

"熊"之初

新生的（newborn）大熊猫是个小不点儿，体长跟一支铅笔差不多，身上只有一层稀疏的白毛。

我已经来到这个世界15天了，但我的外形还是没怎么变。我的皮肤依旧是粉嫩粉嫩的。

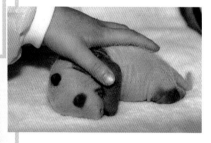

不过，再等5天，你就会看到我身上哪些部位以后会长出黑色毛发了。

饲养员会给我定期称体重。这样，根据我的体重变化，饲养员就会知道妈妈照顾得好不好，我吃饭乖不乖了。

我两个月大喽
(I'm two months old)

成年大熊猫一般独来独往(live alone),只有一种情况例外,那就是熊猫妈妈抚育幼崽(cub)的阶段。熊猫爸爸是从不照顾熊猫宝宝的。

我还不会走路,当然得缠着妈妈了。

小懒虫

跟人类的小宝宝（baby）一样，熊猫幼崽也很能睡。即便已经长大成"熊"，它们每天也有一半的时间在睡觉。

天灵灵，地灵灵，全部没有妈妈灵

晶晶的妈妈娅娅已经不是第一次当妈妈，养儿育女还是很得心应手的。

勇攀新高
（Reaching new heights）

我是爬树（climbing the tree）高手！别看我胳膊（arm）短、腿（leg）也短，可我的力气大着呢！高手都是练出来的，通过不断练习，我现在已经能用爪子（claw）牢牢抓住（grip）树枝，安稳地待在树上了。

"熊"往高处走

这一番攀爬还真挺累的，美美地歇一阵儿吧！大熊猫很多时候都待在树上，这样不会爬树的捕食者（predator）便只有望树兴叹的份儿了。

到了求偶的时节（season），雌性（female）大熊猫有时候会爬到树上，而雄性大熊猫则为了赢得"美人"的芳心在树下大打出手。

吃大餐啦
（Time for dinner）

我现在8个月大，开始进食固体（solid）食物了。学着妈妈的样子，我也吃起了竹子。竹子所含的能量（energy）不多，我只好大吃特吃，用数量弥补质量了。我一天大部分时间都在进食或睡觉。

竹子小知识

- 快速生长期的竹子一天能长高1米，够快吧！
- 竹子对生长环境的要求不高，无论气候炎热还是寒冷，海拔高还是低，竹子都能生长。
- 中国境内的竹子种类多达上千种。

饮水

野生环境中，由于大熊猫的栖息地往往临近河流（river），大熊猫可以随时开怀畅饮。

嗯，好好吃啊，鲜美多汁的竹笋！

竹笋是大熊猫最爱吃的食物。

游戏时间
（Panda's playtime）

我可贪玩儿了！野生环境中，妈妈和我一起游戏。我正是通过这种方式学习生存技能的。成年（adult）后的我们喜欢独来独往，一般不会结伴玩耍。

荡荡秋千吧！

嗨，我在上面呢！

边玩边学

在成都大熊猫繁育研究基地，晶晶有一个木头搭建的攀爬架（climbing frame）。在这个特殊的游乐场（playground）里，熊猫宝宝可以一边玩耍一边练习攀爬了。

大熊猫保护

乐在成都
（Life in Chengdu）

在中国成都，有一个大熊猫繁育研究基地（以下简称基地），这个基地致力于大熊猫（Giant Panda）的保护和研究工作。

成都是四川省的省会，位于我国中部。基地离成都市区不远，建造得犹如大熊猫在山林（forest）中的自然家园。

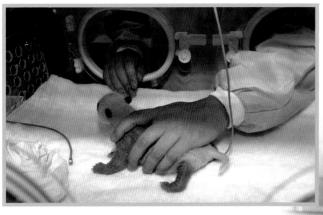

大熊猫繁育管理

生活在基地里的大熊猫有专门的团队（team）照料。熊猫宝宝刚出生时，动物医生就会为它们体检，饲养员则会为幼崽喂食并跟它们玩耍，这都是照料大熊猫的一部分。

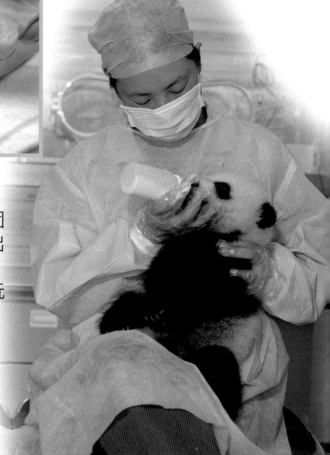

兽舍

成都大熊猫繁育研究基地的兽舍里有攀爬架、水池（pool）以及各种各样的玩具（toy），全都是大熊猫喜欢的东西。成年大熊猫独享一个兽舍，幼年大熊猫有时候则跟同伴共享一个兽舍。

晶晶

晶晶是一个大明星，晶晶于2005年8月在成都出生。2008年北京奥运会的吉祥物之一——晶晶的名字就来源于我们的熊猫晶晶。

年幼的晶晶很贪吃。要来住它的兽舍？好的，欢迎！要它的饲养员照顾你？好的，也没问题！但是，如果要抢它的竹子？哈哈，那可没门儿！

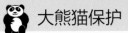 大熊猫保护

生生不息
（Breeding success）

大熊猫属于濒危物种（endangered species）。通过促进大熊猫繁育，成都大熊猫繁育研究基地等机构有效地参与了这一物种的保护工作。

这几只大熊猫都是2006年出生的，同年出生的一共有12只呢！

教育

在成都大熊猫繁育研究基地,游客(visitor)不仅可以增进对大熊猫的了解,还能学到不少关于如何保护动物的知识,学会与动物和谐共处,共同保护我们生活的星球(planet)。

婴儿潮

自2005年以来,共有超过150只大熊猫在基地出生,晶晶便是其中一员。

孕育生命

截至2019年,全球约有600只圈养(captive)大熊猫。它们都是繁育计划的一部分,计划的目的就是帮熊猫繁衍后代。所以,当妈妈也是晶晶的一项光荣使命!

生命循环，周而复始
The circle of life goes round and round

现在你知道我怎样成长为一只成年大熊猫了吧！

Now you know how I turned into a grown-up panda.

词汇表 Glossary

竹子
Bamboo

一种高大的禾本科植物，它的茎十分坚韧，中间是空心的。

皮毛
Fur

动物体表用来御寒、保暖的一层厚实的皮肤和毛发。

爪
Claw

动物脚掌上短而尖利弯曲的硬质趾（指）甲。

饲养员
Keeper

动物园或其他类似场所中负责照顾动物的人。

幼崽
Cub

对刚出生到一岁之前的很多哺乳动物（如熊、狐或狮）的称呼。

捕食者
Predator

猎食其他动物的动物。

致谢 Acknowledgements

感谢以下人员及机构提供图片：

(Key: a=above; c=centre; b=below; l=left; r=right; t=top)
Alamy Images: Steve Bloom Images 1, 6-7; LMR Group 22clb; Keren Su/ China Span 2-3; Andrew Woodley 5tr; Anna Yu 24crb. Ardea: M. Watson 22cb. Chengdu Research Base pandaphoto网站: Zhang Zhihe 4crb, 6cr, 9c, 9ca, 10-11, 11ca, 11cr, 11tl, 12, 13cla, 13clb, 13r, 13tl, 14br, 14cl, 14-15, 16, 16tl, 17r, 17tl, 18br, 18cl, 18tr, 19br, 19clb, 19tl, 20-21, 21cr, 21tl, 22c, 22cr, 22tc, 22tr, 23, 24br, 24cla, 24clb, 24cra, 24tr. Corbis: Brooks Kraft 22crb; Phototex/epa 22cl; Reuters/Henry Romero 22tl. FLPA: Gerry Ellis/Minden Pictures 5cla, 6clb, 8-9, 9br, 9tr. OSF: Mike Powles 4cl. PA Photos: AP Photo/Color China Photo 20cl. Photoshot /NHPA: Gerard Lacz 4-5.
Jacket images: Front: Ardea: M. Watson tl. Getty Images: The Image Bank/Daniele Pellegrini tr; Minden Pictures/Gerry Ellis ftr. Back: Alamy Images: LMR Group clb. Ardea: M. Watson cb. Chengdu: c, cr, tc, tr. Corbis: Brooks Kraft crb; Phototex/epa cl; Reuters/Henry Romero tl. Getty Images: Minden Pictures/Cyril Ruoso/JH Editorial (b/g). Spine: Getty Images: The Image Bank/ Daniele Pellegrini.

其他图片版权属于多林·金德斯利公司。欲了解更多信息请访问DK Images网站。

Glossary

Bamboo
A kind of giant grass with tough, hollow stems.

Fur
The thick, hairy coat that keeps an animal warm.

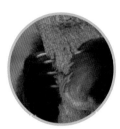
Claw
One of the short, sharp, curved nails on an animal's paw.

Keeper
Someone who looks after animals in a zoo or centre.

Cub
A baby bear (or fox, or lion) in its first year.

Predator
An animal that kills and eats other animals.

Acknowledgements

感谢以下人员及机构提供图片：
(Key: a=above; c=centre; b=below; l=left; r=right; t=top)

Alamy Images: Steve Bloom Images 1, 6-7; LMR Group 22clb; Keren Su/ China Span 2-3; Andrew Woodley 5tr; Anna Yu 24crb. Ardea: M. Watson 22cb. Chengdu Research Base pandaphoto网站: Zhang Zhihe 4crb, 6cr, 9c, 9ca, 10-11, 11ca, 11cr, 11tl, 12, 13cla, 13clb, 13r, 13tl, 14br, 14cl, 14-15, 16, 16tl, 17r, 17tl, 18br, 18cl, 18tr, 19br, 19clb, 19tl, 20-21, 21cr, 21tl, 22c, 22cr, 22tc, 22tr, 23, 24br, 24cla, 24clb, 24cra, 24tr. Corbis: Brooks Kraft 22crb; Phototex/epa 22cl; Reuters/Henry Romero 22tl. FLPA: Gerry Ellis/Minden Pictures 5cla, 6clb, 8-9, 9br, 9tr. OSF: Mike Powles 4cl. PA Photos: AP Photo/Color China Photo 20cl. Photoshot /NHPA: Gerard Lacz 4-5.
Jacket images: Front: Ardea: M. Watson tl. Getty Images: The Image Bank/Daniele Pellegrini tr; Minden Pictures/Gerry Ellis ftr. Back: Alamy Images: LMR Group clb. Ardea: M. Watson cb. Chengdu: c, cr, tc, tr. Corbis: Brooks Kraft crb; Phototex/epa cl; Reuters/Henry Romero tl. Getty Images: Minden Pictures/Cyril Ruoso/JH Editorial (b/g). Spine: Getty Images: The Image Bank/Daniele Pellegrini.

其他图片版权属于多林·金德斯利公司。欲了解更多信息请访问DK Images网站。

The circle of life goes round and round

Now you know how I turned into a grown-up panda.

Education

In Chengdu, visitors learn about pandas, and also about conservation – how to look after our planet and the animals we share it with.

Baby boom

Jingjing is one of 150 pandas that have been born in Chengdu since the year 2005... so far!

A job for life

By the end of 2019, there are about 600 captive pandas in the world. They are all part of a breeding programme to help them have cubs. Being a mother will become Jingjing's job too!

PANDA CONSERVATION

Breeding success

Giant pandas are an endangered species. There are fewer than 1,900 left in the world. Places like Chengdu help to conserve (or save) the species by helping their pandas breed.

These are some of the 12 pandas born in 2006.

Enclosures

In Chengdu, a panda's enclosure is full of things a panda loves, such as climbing frames, a pool, and toys. Adult pandas live alone, but young pandas sometimes share.

Jingjing

The star of this book, Jingjing, was born at Chengdu in August 2005.
She has been chosen as one of the mascots of the Beijing Olympics.

Jingjing was just over two years old when this book was made. She shares her enclosure, and her keepers, with two other young pandas – but not her bamboo!

PANDA CONSERVATION

Life in Chengdu

The Chengdu Research Base of Giant Panda Breeding in China works to save giant pandas. Today there are 47 pandas being looked after at the base.

Chengdu is in central China. The base is near a city, but was built to look like a panda's natural forest home.

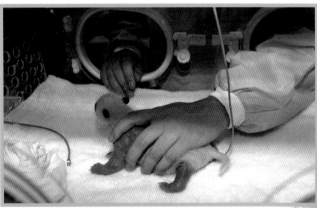

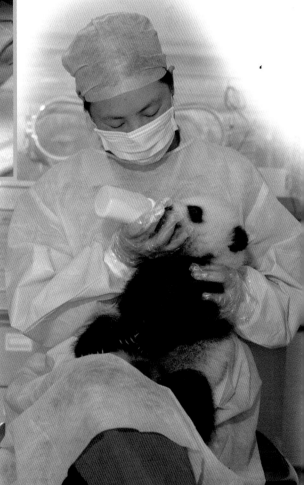

Looking after the pandas

Pandas at Chengdu have a whole team of staff to help care for them. Vets check their health from the moment they are born, and keepers feed and play with cubs as part of looking after them.

I'm up here now!

Play and learn

At the Chengdu Panda Base, Jingjing has a bamboo climbing frame. The cubs in Chengdu are given these special playgrounds so they can practice climbing while they play.

Panda's playtime

Pandas love to play! In the wild, mums and cubs play together. This is how the cub learns. Adult pandas live and play alone.

Let's try the swing

Yum! Sweet, juicy bamboo!

Bamboo shoots are the pandas' favourite food.

Time for dinner

Now I'm eight months old, I am starting to explore solid food. My main food is bamboo, which I learn to eat by copying mum. Bamboo doesn't give me much energy, so I need to eat a lot of it. I spend most of my day eating and resting.

Bamboo facts

- Bamboo can grow up to 1 m (3 ft) a day. That's fast!
- Bamboo can grow in hot places, cold places, high places, and low places.
- There are thousands of different species of bamboo in China.

Drinking water

In the wild, pandas often live near rivers so they have lots of water to drink.

In the mating season, female pandas sometimes climb a tree while males fight over her on the ground.

Reaching new heights

I'm an expert in climbing trees! My arms and legs are short, but they are strong. It took a lot of practice, but now I can pull myself up and hang on. I grip the branches with my sharp claws.

A place to think
After all that climbing, it's good to stop and rest. Pandas spend a lot of time up in trees, where they're safe from predators.

Sleepy head
Panda cubs sleep a lot of the time, just like human babies. Even fully grown pandas spend nearly half their day resting.

Mum's the word
Yaya, Jingjing's mum, has other cubs who are older than Jingjing, so Yaya knows all about how to be a mum.

I'm two months old

Adult pandas live alone. The only time an adult lives with another panda is when a mum lives with her cub. Panda dads don't get involved.

I can't walk yet, so I stay very close to mum.

Panda facts

- Panda mums usually give birth to one or two cubs at a time.
- Cubs stay with their mum for at least 18 months. Some stay longer.
- It takes about 5 years for a cub to become an adult.

Nothing beats a cuddle from mum!

Early days

A newborn panda cub is tiny – about as long as a pencil! It's born with a little white fur.

At 15 days old, I haven't changed much. My skin is still mostly pink.

... But about five days later, you can see where my black fur will grow.

I get weighed as I grow. This way, my keepers can be sure mum is looking after me and I'm eating well.

This is my mum

Jingjing's mother is named Yaya. When she gave birth to Jingjing, her cub was taken to be weighed and checked, but was soon given back to mum to be looked after.

This is my home

Pandas live in China, in the mountains where bamboo grows. In the wild, they don't live anywhere else in the world. Bamboo is the panda's favourite food.

A different home

Jingjing was born in Chengdu Panda Base, in China. Her mother and father live there too. She has keepers to help look after her.

Time for a rest

Pandas find resting places in bamboo groves or rock dens. Young pandas rest in trees.

Winter in the mountains can be very cold and snowy.

Furry paws help pandas to walk on slippery snow and steep rocks.

Their teeth are strong enough to bite through thick bamboo stems.

I'm a giant panda

I'm a bear that lives in China. I have thick black and white fur. I spend half my day eating bamboo. The rest of the time I like to sleep and play – just like you!

Super senses
Pandas have a good sense of smell. Their hearing is excellent too, which is why they're hard to find – when they hear people coming, they leave!

Pandas have thick, oily fur to keep out the cold and wet weather.

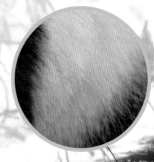

Now turn the page and see how I grow...

12~13
Reaching new heights

14~15
Time for dinner

16~17
Panda's playtime

18~19
Life in Chengdu

20~21
Breeding success

22~23
The circle of life

24
Glossary

Contents

4~5
I'm a giant panda

6~7
This is my home

8~9
This is my mum

10~11
I'm two months old

Contents

4~5
I'm a duck

6~7
Before I was born

8~10
Inside my egg I'm warm and snug

11
It's time for me to break out

12~13
I'm two days old

14~15
I'm off for my first swim

16~17
I'm four weeks old

18~19
I'm ready to fly

20~21
The circle of life

22~23
My friends from around the world

24
Glossary

I'm a duck

I'm a great swimmer and I can fly too! My body is covered with soft, oily feathers. I am completely waterproof.

Look closely
Can you see the water droplets rolling off the duck's oily feathers?

Ducks have soft feathers to keep them warm.

Ducks have a stretchy web of skin in between each long toe.

Ridges on a duck's bill help it to grip its food.

Turn the page and find out how my life began.

Bottoms up
Some ducks dip underwater to search for food. This is called dabbling.

Gobble... Gurgle... Slurp...

Before I was born

Mum and dad met in the spring and built a nest. Soon mum laid her eggs and I was in one of them.

Together
Male and female ducks stay together until their eggs are laid.

Bright and beautiful
Male ducks are very colourful. They attract female ducks by ruffling their own feathers.

This is my dad.

Under cover
Female ducks are the same colour as the nest. This helps them hide their eggs.

This is my mum.

Nest making
Female ducks use feathers, twigs, and grasses to build a cosy, warm nest.

Inside my egg I'm warm and snug

My mum sits on our eggs to keep them warm. She sits on them night and day until it's time for us to hatch.

Feather blanket
The nest is lined with feathers to make it warm and soft. Inside the nest, the eggs are safe and snug.

Watch out - danger about!

Duck eggs are a tasty treat for many animals, such as rats and foxes. The mother duck keeps a sharp lookout for hungry animals. When she spots one, she protects her eggs by chasing it away.

Rat

Fox

Tap... Tap... Crack... Crack...

Look at this egg. Can you see a tiny, pink beak?

Tap, tap, peek-a-boo toes.

Tap, tap, wait for me!

It's time for me to break out

Inside the shell, I start to peck. It takes me about 10 hours to peck my way out of my shell.

Puff, pant, I'm nearly there.

Phew! Done it!

I'm two days old

It's time to go for my first waddle. It's hard work. I make sure I stay close to mum, so that she can look after me.

My brothers, sisters, and I have to run to keep up with mum.

Mother ducks watch for any signs of danger.

She likes to take such big, strong steps!

Walkabout facts

- Ducks have no nerves in their feet so their toes never get cold, even on icy ponds.
- Ducks' feet are made of the same bendy stuff that is in your nose and ears.

I'm off for my first swim

After three days I make my way to the water, where I start to paddle. My feathers are waterproof, and they keep me warm and dry.

Did you know?
· · · · · · · · · · · · · · · · · ·
When baby ducklings get cold, they huddle close to their mum to warm up.

Paddle power
Duckling's flat feet push them along in the water.

Paddle hard my little ones!

Two weeks old
The duckling's beak is growing longer and it has lost all of its fluffy yellow feathers.

I'm four weeks old

It's fun to catch my own meals by dipping under the water. My flat beak helps me scoop up lots of food.

A duckling cannot fly because its wings are not fully grown yet.

Yummy dinner
Guess what? This is a duck's dinner! Millions of tiny insects and tasty green plants are floating around in the pond.

Webbed feet help ducks to balance in the water.

Gobble facts

🦆 Ducks are mostly vegetarian. They eat mainly plants and seeds that grow in or near the water.

🦆 Foods like bread and biscuits are bad for ducks because they swell up in their stomach. Ouch!

I'm ready to fly

I'm eight weeks old and my wings are fully grown. I can't wait to fly with the other ducks. But first I have to learn how to take off and land.

When they take off, ducks tilt their wings and flap very fast.

A duck's tail feathers help it to steer.

Up, Up, and away – I feel as light as a feather!

A final push and the duck lifts into the air.

Flying together

When they travel long distances, ducks always fly with other ducks. A group of ducks is called a flock.

The circle of life goes round and round

Now you know how I turned into a fluffy duck.

My friends from around the world

This is a White-Faced Duck from South Africa.

American Wood Ducks like to build their nests in trees close to water.

Pekin Ducks have fluffy yellow chicks.

The Black-Bellied Whistling Duck makes a sound like "pe-che-che".

There are hundreds of different kinds of ducks in the world. Are there any ducks living near you?

Mandarin Ducks live in China and like to eat rice and other grains.

The Shoveler Duck uses its big bill to dig for food.

phwee-eek

The Plumed Whistler makes a sound just like a squeaky whistle.

Fun duck facts

- The Black Brant duck can fly more than 1,600 kilometres (960 miles) without stopping.
- Most ducks fly south in autumn, so they can spend the winter in a warm place.
- Male ducks are usually more colourful than female ducks.

Glossary

Feather
Soft, light parts that cover the outside of a bird's body.

Hatching
When a baby duck or other animal comes out of its egg.

Waterproof
Something that does not let water pass through it.

Nest
A place that a duck builds out of twigs to lay its eggs in.

Webbing
The thin skin between a duck's toes.

Dabbling
When ducks feed by sticking their heads under water.

Acknowledgements

感谢以下人员及机构提供图片：
(Key: a=above; c=centre; b=below; l=left; r=right; t=top)
1: Windrush Photos/Colin Carver c; 2: Masterfile UK c; 2-3: N.H.P.A./ Manfred Danegger b; 3: Ardea London Ltd/Tom & Pat Leeson tr; 4: Ardea London Ltd/John Daniels bl; Pat Morris tr; 4-5: Global PhotoSite Copyright © 2000 - 2003 Chris Edwards/Mal Smith cl; 5: Global PhotoSite Copyright © 2000 - 2003 Chris Edwards/ Mal Smith tr; 5: Oxford Scientific Films br; 6: Ardea London Ltd/ Kenneth W. Fink tr; 6-7: Holt Studios/Primrose Peacock; 7: Oxford Scientific Films/ Mark Hamblin br; 8: Premaphotos Wildlife/KG Preston-Mafham l; 8-9: FLPA - Images of Nature/Maurice Walker; 10-11: Premaphotos Wildlife/K G Preston-Mafham c; 12-13: Oxford Scientific Films/ Wendy Shattil and Bob Rozinski;

14-15: Ardea London Ltd/Brian Bevan, 15: Ardea London Ltd/John Daniels tr; 15: FLPA - Images of nature/Tony Wharton br; 16: Ardea London Ltd/John Daniels l; 17: Ardea London Ltd/John Daniels; 18: Ardea London Ltd/John Daniels; 19: Ardea London Ltd/Chris Knights c; John Daniels bl; 20: Ardea London Ltd/John Daniels tl, br, bra; 20: Chris Gomersall Photography cl; 20: Holt Studios/Wayne Hutchinson c; 21: FLPA - Images of nature/Jurgen & Christine Sohns; 22: Ardea London Ltd/Jim Zipp 2000 bl; Kenneth W. Fink cla; 22: Holt Studios/ Mike Lane tr; 23: Ardea London Ltd/ Kenneth W. Fink tr; 23: Getty Images/Richard Coomber tl; 23: Masterfile UK br; 24: Ardea London Ltd/John Daniels bl; Pat Morris cl; 24: Windrush Photos/David Tipling br. Jacket Front: Barrie Watts tcr, Ardea/John Daniels bc.

其他图片版权属于多林·金德斯利公司。欲了解更多信息请访问DK Images网站。

词汇表 Glossary

羽毛 Feather
披覆在鸟儿全身，又轻又软的部分。

孵化 Hatching
指鸭宝宝或其他卵生动物从蛋里破壳而出。

防水 Waterproof
某些材料具有的阻隔水的特性。

巢 Nest
用小树枝搭建、鸭妈妈用来孵蛋的地方。

鸭蹼 Webbing
鸭子脚趾之间薄薄的皮肤。

水中觅食 Dabbling
指鸭子将头扎入水中寻找食物。

致谢 Acknowledgements

感谢以下人员及机构提供图片：

(Key: a=above; c=centre; b=below; l=left; r=right; t=top)
1: Windrush Photos/Colin Carver c; 2: Masterfile UK c; 2-3: N.H.P.A./Manfred Danegger b; 3: Ardea London Ltd/Tom & Pat Leeson tr; 4: Ardea London Ltd/John Daniels bl; Pat Morris tr; 4-5: Global PhotoSite Copyright © 2000 - 2003 Chris Edwards/Mal Smith cl; 5: Global PhotoSite Copyright © 2000 - 2003 Chris Edwards/ Mal Smith tr; 5: Oxford Scientific Films br; 6: Ardea London Ltd/ Kenneth W. Fink tr; 6-7: Holt Studios/Primrose Peacock; 7: Oxford Scientific Films/ Mark Hamblin br; 8: Premaphotos Wildlife/KG Preston-Mafham l; 8-9: FLPA - Images of Nature/Maurice Walker; 10-11: Premaphotos Wildlife/K G Preston-Mafham c; 12-13: Oxford Scientific Films/Wendy Shattil and Bob Rozinski; 14-15: Ardea London Ltd/Brian Bevan, 15: Ardea London Ltd/John Daniels tr; 15: FLPA - Images of nature/Tony Wharton br; 16: Ardea London Ltd/John Daniels l; 17: Ardea London Ltd/John Daniels; 18: Ardea London Ltd/John Daniels; 19: Ardea London Ltd/Chris Knights c; John Daniels bl; 20: Ardea London Ltd/John Daniels tl, br, bra; 20: Chris Gomersall Photography cl; 20: Holt Studios/Wayne Hutchinson c; 21: FLPA - Images of nature/Jurgen & Christine Sohns; 22: Ardea London Ltd/Jim Zipp 2000 bl; Kenneth W. Fink cla; 22: Holt Studios/ Mike Lane tr; 23: Ardea London Ltd/ Kenneth W. Fink tr; 23: Getty Images/Richard Coomber tl; 23: Masterfile UK br; 24: Ardea London Ltd/John Daniels bl; Pat Morris cl; 24: Windrush Photos/David Tipling br. Jacket Front: Barrie Watts tcr, Ardea/John Daniels bc.

其他图片版权属于多林·金德斯利公司。欲了解更多信息请访问DK Images网站。

世界（world）上有几百种不同的鸭子。
你住的地方附近有没有鸭子呢？

鸳鸯（Mandarin Duck）生活在中国，喜欢吃小虫和各种谷物。

琵嘴鸭（Shoveler Duck）靠自己宽大的喙刨出食物。

鸭子趣闻

- 🦆 黑雁鸭能够不间断地飞行超过1600千米。

- 🦆 到了秋天，北半球的大多数鸭子都要往南飞，这样就可以在温暖的地方过冬了。

- 🦆 雄鸭的羽毛颜色往往比雌鸭的更艳丽多彩。

尖羽树鸭（Plumed Whistler）可以发出像尖锐的哨声一样的声音。

我世界各地的朋友
My friends from around the world

这是南非的白脸树鸭（White-Faced Duck）。

林鸳鸯（American Wood Duck）喜欢在临近水边的树上筑巢。

北京鸭（Pekin Duck）的宝宝是黄色的，毛茸茸的。

黑腹树鸭（Black-Bellied Whistling Duck）会发出奇特的叫声。

嘎嘎嘎,我要去撩撩小伙伴的羽毛啦!

生命循环，周而复始
The circle of life goes round and round

现在你知道我怎样长成
一只毛茸茸的鸭子了吧！

Now you know how I turned into a fluffy duck.

飞啊,飞啊,飞起来啦!
我觉得自己就像羽毛一样轻盈!

奋力一蹬,冲上云霄。

一起飞翔

长途飞行时,我们总是会结伴同行,形成一支"队伍"。

准备起飞
（I'm ready to fly）

我已经8周大（eight weeks old）了，翅膀也发育完全了，我等不及要和大家一起飞上蓝天啦！但是在这之前，我必须先学会如何起飞（take off）和降落（land）。

起飞时，我们的翅膀会稍稍倾斜（tilt）并快速地拍打（flap）。

尾羽（tail feather）有助于我们在飞行时控制方向。

脚蹼（webbed feet）能帮助我在水中保持平衡（balance）。

鸭子小知识

- 鸭子一般以素食为主，主要的食物就是水域附近的植物和种子。
- 千万别喂鸭子吃面包和饼干，因为它们会在鸭子的胃里膨胀起来。那样鸭子会很难受！

我4周大了
（I'm four weeks old）

潜水觅食真是太有趣了！扁扁的嘴巴能帮（help）我捕捉到很多食物。

我的翅膀（wing）还没完全发育好，所以现在我还飞不起来。

美味的晚餐

猜猜看，这是什么？这些漂浮在池塘（pond）水面上的小昆虫（insect）和绿色植物（green plant）就是我美味的晚餐（dinner）！

划水的力量

游泳时,我们用扁平的脚蹼推动(push)自己前进。

小家伙们,用力划啊!

出生两周了

我的嘴巴越长越长。出生时那身蓬松的黄色(yellow)绒毛已经完全脱落了。

第一次下水
（I'm off for my first swim）

出生3天后（after three days），我来到水边学习划水。感谢这身防水的（waterproof）羽毛，让我的身体保持温暖与干燥。

你知道吗？

每当我们感到寒冷（cold）时，就会聚拢到妈妈身边取暖。

妈妈会留意任何危险的迹象。

妈妈走起路来总是昂首阔步、坚定有力!

鸭掌小知识

🦆 鸭掌上没有神经,所以即便是走在结冰的池塘上,它们的脚趾也不会觉得冷。

🦆 鸭掌中的软骨与人类鼻子和耳朵中的软骨的成分是相似的。

我出生两天了
(I'm two days old)

这是我第一次（first）走路，真的好难啊！我一定要跟紧妈妈，这样她才能照顾（look after）到我。

我和兄弟姐妹得一路小跑才能跟得上妈妈。

是时候破壳而出了
（It's time for me to break out）

我在蛋壳里啄（peck）呀啄，大约啄了10小时后，我终于破壳而出啦！

咔嚓，咔嚓，摩擦摩擦，我来啦！

加油！就快成功了。

啊！总算出来啦！

小心,有危险!

对老鼠(rat)和狐狸(fox)这样的动物(animal)来说,鸭蛋是可口的美食,所以鸭妈妈对这些饥饿(hungry)的家伙总是保持着高度警惕。一旦察觉到周围有危险,它就会立刻冲上去赶走它们。

老鼠

狐狸

我在蛋里，温暖又舒适
（Inside my egg I'm warm and snug）

为了给我们保暖（warm），妈妈不分昼夜地卧在上面，直到我们破壳而出！

羽毛毯

巢里铺了很多羽毛，像一床温暖又柔软的毛毯一样。我们在里面既安全（safe）又舒适（snug）。

天然保护色

鸭妈妈的毛色和鸭巢的颜色非常接近，这样有助于鸭妈妈把蛋（egg）隐藏起来。

这是我妈妈。

筑巢

鸭妈妈用羽毛、树枝和小草筑起了一个温暖的家。

我出生之前
（Before I was born）

爸爸和妈妈在春天（spring）相遇并筑巢。很快，妈妈就产下了许多鸭蛋，而我，就是其中一员。爸爸的羽毛色彩鲜艳，漂亮非凡。妈妈就是因此而被它吸引的呢！

在一起

在我出生之前，爸爸妈妈形影不离。

这是我爸爸。

我的嘴巴边缘有一条凸出的齿状结构，能帮助我牢牢地咬住食物（food）。

翻开下一页，看看我是怎样长大的吧！

水下觅食

我常常会头朝下、脚朝上地潜到水面下寻觅（search）食物。

一口吞下……
咕噜噜……
嘟嘟嘟……

我是一只鸭子
（I'm a duck）

我不仅是一名游泳健将（swimmer），而且还会飞（fly）呢！我的身上覆盖着柔软且油亮的羽毛，它们可是超级防水的哟！

观察

你能看见水珠（water droplet）从我光滑、油亮的羽毛上滑落吗？

我的羽毛（feather）蓬松而柔软，能帮助我保持温暖。

我的每个长脚趾（toe）之间都有一块富有弹性的皮肤（skin），叫作"蹼"。

目录 Contents

4~5
我是一只鸭子

6~7
我出生之前

8~10
我在蛋里,温暖又舒适

11
是时候破壳而出了

12~13
我出生两天了

14~15
第一次下水

16~17
我4周大了

18~19
准备起飞

20~21
生命循环,周而复始

22~23
我世界各地的朋友

24
词汇表

Original Title: Duckling
Copyright © Dorling Kindersley Limited, 2003
A Penguin Random House Company

本书简体中文版授权由人民邮电出版社独家出版，仅限于中国境内（不包括香港、澳门、台湾地区）销售。未经出版者书面许可，不得以任何方式复制或发行本书中的任何部分。

跟我们一起飞翔，看着我们成长吧！

FSC 混合产品 源自负责任的 森林资源的纸张 FSC® C018179

For the curious
www.dk.com

DK 动物成长奥秘
看！我在长大（中英双语版）

鸭子

英国 DK 公司◎编

钟子悠 朱语行 赵泰然 蔡至钧◎译

鹰之舞 沈成◎审

人民邮电出版社

北京

DK WATCH ME GROW
PENGUIN

Contents

4~5
I'm a penguin

6~7
My dad and mum

8~9
Dad is left in charge

10~11
It's time for me to hatch

12~13
My dad and mum both look after me

14~15
My adult feathers grow in

16~17
My first swim

18~19
My first fishing lesson

20~21
The circle of life

22~23
My friends from around the world

24
Glossary

Come dive with us and watch us GROW!

I'm a penguin

I live in the Antarctic where it is very cold. I am a bird but I do not fly. Instead, I swim in the ocean to find my food.

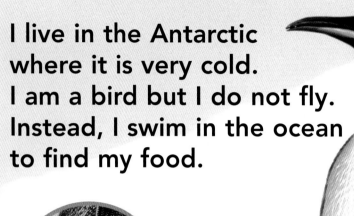

Penguins use their flippers like wings to help them move under the water.

Long, sharp claws help the penguin to grip the ice and snow.

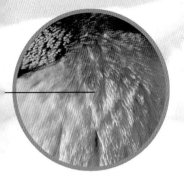

The penguin's ears are underneath this orange and yellow fur.

All together now

When they are on land, these Emperor penguins live together in a big group called a colony.

Turn the page and find out how our lives began...

My dad and mum

My dad and mum spend most of the year apart at sea. In April, when it is time to mate, they come onto land. They recognize each other by their cries.

Penguin facts

- Penguins return to the same colony every year.
- There are only about 40 colonies of Emperor penguins in the world, and about 400,000 adults.
- The adult penguins are about the height of a 3 or 4-year-old child.

Making noise

Emperor penguins have very loud voices. This is important, since they recognize each other by their voice, not by sight.

This is my dad.

Eeearreekkk...

This is my mum.

Dad is left in charge

After mum lays my egg, she gives it to dad, who tucks it under his pouch. Now it's time for mum to go off and find food while dad keeps my egg warm and safe.

Father penguins don't eat while the mothers are away.

Feather bed
Look closely at this picture. Can you see the egg sitting snugly under its dad's fluffy feathers?

Standing still

Father penguins have to move slowly and carefully so they don't drop their eggs.

It takes mum days to reach the sea.

Foot-powered travel

Female penguins scoot off to the sea by sliding on their tummies, since this is quicker than walking. They spend the winter feeding in the sea, then return in the spring just before their egg hatches.

It's time for me to hatch

Dad has been keeping me warm for two months. I will be born in his pouch. When mum returns, he will give me to her and go to eat.

Crack, crack, crack...

...here I come.

Breaking out
The chick uses a hard spot on its beak, called an egg tooth, to peck its way slowly out of the egg. It can take three days to hatch out.

Free at last!

Feather care
The chick's downy feathers are soft and warm, but they are not waterproof. Mum helps keep them clean and fluffy.

I'm two days old.

Mum's big toes lift the chick off the frozen ground and keep it warm.

My dad and mum both look after me

After I am born, mum and dad take turns feeding me and keeping me safe and warm.

Parents recognize their chick by its cry.

Spitting up dinner
The penguin parents store food in their stomachs and then spit it up for the chick.

While their parents gather food, the chicks huddle together in groups called crèches.

Chick facts

- The chicks only eat about 16 meals in the five months it takes them to grow up.
- Huddling together keeps the chicks warm.

My adult feathers grow in

After five months, I am starting to grow my adult, waterproof feathers. This means I'll be able to swim and find my own food.

Squawk squaaark!! Feed me!

This penguin is as big as his father, but he can't feed himself yet.

Waterproofing

It takes the penguin a few weeks to lose its baby feathers and grow adult feathers. This is called fledging.

All change

Adult penguins also grow new feathers each year as the old ones wear out.

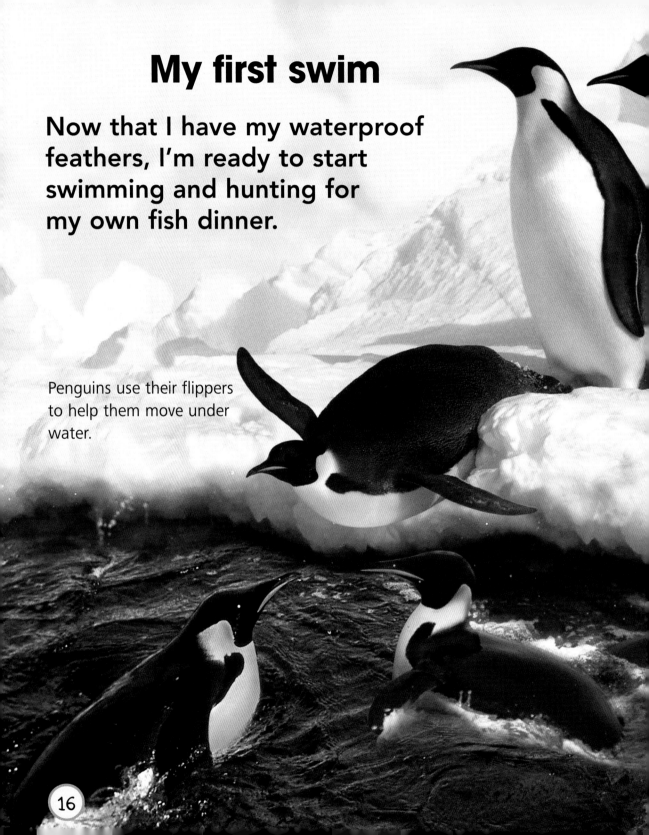

My first swim

Now that I have my waterproof feathers, I'm ready to start swimming and hunting for my own fish dinner.

Penguins use their flippers to help them move under water.

These penguins are waiting in line to dive into the icy cold water.

Penguin dinner

Penguins eat many kinds of food, like fish and shellfish, but their favourite foods are squid and small shrimp called krill.

My first fishing lesson

It's time for me to learn how to dive and how to catch tasty fish to eat. I learn how to catch my own dinner by watching the adult penguins.

The penguins follow shoals of fish. When they catch a fish, they swallow it whole.

A flying finish

The only time that Emperor penguins fly is when they leave the water. They leap out and make a belly flop onto the ice.

These mackerel are a tasty meal.

Emperor penguins can spend up to 20 minutes underwater in one breath.

The circle of life goes round and round

Now you know how I turned into a grown-up penguin.

Bye bye, see you next winter.

My friends from around the world

This Little Blue Penguin is only 25 cm (12 in) tall.

The Gentoo Penguin is the only penguin that raises two chicks at one time.

The Humbolt Penguin lives in warm countries such as South Africa.

My favourite food is squid.

This is a Chinstrap Penguin. Can you see how it got its name? It lives on an Antarctic island.

My friends and I all have black and white feathers, and some of us have amazing hairstyles too.

The Macaroni Penguin is the most common penguin.

Rockhopper Penguins move by hopping from rock to rock.

I am about 1 metre (3 ft) tall.

I'm a Yellow-Eyed Penguin.

King Penguins look a lot like Emperor Penguins, but they are smaller and have more orange colour.

Penguin facts

- Penguins can drink both saltwater and freshwater.
- Penguins have bristles on their tongues that keep slippery seafood from getting away.
- Penguins only live in the Southern hemisphere.

Glossary

Flippers
The wings of the penguin. They help the penguin swim in water.

Pouch
A flap of fur the baby penguin snuggles under to keep warm.

Colony
A group of penguins that lives together in one place.

Crèche
A group of baby penguins that is looked after by one adult.

Hatch
When the baby penguin pecks its way out of its egg.

Fledge
When a penguin grows its adult, waterproof feathers.

Acknowledgements
感谢以下人员及机构提供图片：
(Key: a=above; c=centre; b=below; l=left; r=right; t=top)
1 Corbis: Tim Davis bkgnd; Royalty Free Images: Corbis. 2-3 Getty Images: Tim Davis. 2 Corbis: Kevin Schafer tr. 3 Corbis: Wolfgang Kaehler bl; Zefa Picture Library: H. Reinhard c. 4-5 Corbis: Tim Davis. 4 Bruce Coleman Ltd: Johnny Johnson cl; FLPA - Images of Nature: F Lanting/ Minden Pictures bc. Getty Images: Pete Oxford r. 5 Alamy Images: Fritz Poelking/ Elvele Images tr. Corbis: Clive Druett tl. 6-7 ImageState/Pictor: David Tipling c, t; Corbis: Tim Davis b. 7 Alamy Images: Galen Rowell/ Mountain Light. 8-9 FLPA - Images of Nature: Frans Lanting. 8 Alamy Images: Paul Gunning l; Nature Picture Library Ltd: Doug Allan br. 9 Science Photo Library: Doug Allan t. 10-11 FLPA - Images of Nature: Frans Lanting/ Minden Pictures. 10 DK Images: Jane Burton tl, cr; Sea World/ San Diego Zoo: br. 12-13 FLPA - Images of Nature: Frans Lanting/ Minden Pictures. 12 Bruce Coleman Ltd: Hans Reinhard c; FLPA – Images of Nature: K Wothe/ Minden Pictures l. 14-15 Corbis: John Conrad. 15 ImageState/Pictor: Hummel/ Fotonatura; FLPA - Images of nature: C Carvalho tr. 16 Bruce Coleman Ltd: Johnny Johnson bl; Corbis: Tim Davis tl;
Robert Harding Picture Library: Johnny Johnson br. 17 Corbis: Tim Davis t; DK Images: Frank Greenaway bc; Harry Taylor br. 18-19 Corbis: Stuart Westmorland. 18 National Geographic Image Collection: Bill Curtsinger. 19 Nature Picture Library Ltd: Doug Allan c; Pete Oxford tr; Oxford Scientific Films: Doug Allan bkg. 20 Bruce Coleman Ltd: Dr. Eckart Pott tl; Johnny Johnson c; Corbis: Tim Davis clb, crb, bcr; DK Images: Jane Burton tc, tr; Oxford Scientific Films: Doug Allan bcl; Royalty Free Images: Corbis cr; Sea World/ San Diego Zoo: trb; Getty Images: Pete Oxford cr. 21 N.H.P.A.: B & C Alexander. 22-23 FLPA – Images of Nature: Gerald Lacz. 22 Howard Porter: Daniel Zupanc tr; FLPA - Images of Nature: David Hosking bl; Tui De Roy/ Minden Pictures b; Oxford Scientific Films: Gerald L. Kooyman tr. 23 Alamy Images: Bryan & Cherry Alexander br; Ardea London Ltd: cr; Peter Steyn tr; FLPA - Images of Nature: Terry Andrewartha tr. 24 Bruce Coleman Ltd: Johnny Johnson tl; Corbis: Fritz Polking/ Frank Lane Picture Agency cr; ImageState/ Pictor: Hummel/ Fotonatura br; Oxford Scientific Films: Doug Allan tr, cr; Sea World/ San Diego Zoo: bl.

其他图片版权属于多林·金德斯利公司。欲了解更多信息请访问DK Images网站。

词汇表 Glossary

鳍状肢 Flippers
如企鹅翅膀，帮助其在水里游动。

育儿袋 Pouch
企鹅腹部的一块皮肤皱褶，可以给蛋或小企鹅保暖。

族群 Colony
在一个地方一起生活的一群生物，如企鹅。

幼儿园 Crèche
由一只成年企鹅照顾一群企鹅宝宝。

孵化 Hatch
啄开蛋壳出生的过程。

换羽 Fledge
褪去绒毛，长出防水羽毛的过程。

致谢 Acknowledgements

感谢以下人员及机构提供图片：

(Key: a=above; c=centre; b=below; l=left; r=right; t=top)
1 Corbis: Tim Davis bkgnd; Royalty Free Images: Corbis. 2-3 Getty Images: Tim Davis. 2 Corbis: Kevin Schafer tr. 3 Corbis: Wolfgang Kaehler bl; Zefa Picture Library: H. Reinhard c. 4-5 Corbis: Tim Davis. 4 Bruce Coleman Ltd: Johnny Johnson cl; FLPA - Images of Nature: F Lanting/ Minden Pictures bc. Getty Images: Pete Oxford r. 5 Alamy Images: Fritz Poelking/ Elvele Images tr. Corbis: Clive Druett tl. 6-7 ImageState/Pictor: David Tipling c, t; Corbis: Tim Davis b. 7 Alamy Images: Galen Rowell/ Mountain Light. 8-9 FLPA - Images of Nature: Frans Lanting. 8 Alamy Images: Paul Gunning l; Nature Picture Library Ltd: Doug Allan br. 9 Science Photo Library: Doug Allan t. 10-11 FLPA - Images of Nature: Frans Lanting/ Minden Pictures. 10 DK Images: Jane Burton tl, cr; Sea World/ San Diego Zoo: br. 12-13 FLPA - Images of Nature: Frans Lanting/ Minden Pictures. 12 Bruce Coleman Ltd: Hans Reinhard c; FLPA – Images of Nature: K Wothe/ Minden Pictures l. 14-15 Corbis: John Conrad. 15 ImageState/Pictor: Hummel/ Fotonatura; FLPA - Images of nature: C Carvalho tr. 16 Bruce Coleman Ltd: Johnny Johnson bl; Corbis: Tim Davis tl; Robert Harding Picture Library: Johnny Johnson br. 17 Corbis: Tim Davis t; DK Images: Frank Greenaway bc; Harry Taylor br. 18-19 Corbis: Stuart Westmorland. 18 National Geographic Image Collection: Bill Curtsinger. 19 Nature Picture Library Ltd: Doug Allan c; Pete Oxford tr; Oxford Scientific Films: Doug Allan bkg. 20 Bruce Coleman Ltd: Dr. Eckart Pott tl; Johnny Johnson c; Corbis: Tim Davis clb, crb, bcr; DK Images: Jane Burton tc, tr; Oxford Scientific Films: Doug Allan bcl; Royalty Free Images: Corbis cr; Sea World/ San Diego Zoo: trb; Getty Images: Pete Oxford cr. 21 N.H.P.A.: B & C Alexander. 22-23 FLPA – Images of Nature: Gerald Lacz. 22 Howard Porter: Daniel Zupanc tr; FLPA - Images of Nature: David Hosking bl; Tui De Roy/ Minden Pictures b; Oxford Scientific Films: Gerald L. Kooyman tr. 23 Alamy Images: Bryan & Cherry Alexander br; Ardea London Ltd: cr; Peter Steyn tr; FLPA - Images of Nature: Terry Andrewartha tr. 24 Bruce Coleman Ltd: Johnny Johnson tl; Corbis: Fritz Polking/ Frank Lane Picture Agency cr; ImageState/Pictor: Hummel/ Fotonatura br; Oxford Scientific Films: Doug Allan tr, cr; Sea World/ San Diego Zoo: bl.

其他图片版权属于多林·金德斯利公司。欲了解更多信息请访问DK Images网站。

我和我的朋友都有一身黑白相间的羽毛,我们中有些朋友还有着奇特的发型(**hairstyle**)。

马可罗尼企鹅(Macaroni Penguin,别名:长眉企鹅)是最常见的企鹅。

凤头黄眉企鹅(Rockhopper Penguin,别名:跳岩企鹅)喜欢在岩石间跳跃。

我的身高约有1米。

我是一只黄眼企鹅。

王企鹅(King Penguin)和帝企鹅的外形很像,但王企鹅的体形更小,且身上橘色的毛更多。

企鹅小知识

- 企鹅既喝海水也喝淡水。
- 企鹅的舌头上有倒刺,可以防止滑溜溜的鱼虾等食物逃走。
- 企鹅只生活在南半球。

我世界各地的朋友
My friends from around the world

这只小蓝企鹅（Little Blue Penguin）的身高只有25厘米。

巴布亚企鹅（Gentoo Penguin，别名：白眉企鹅、金图企鹅）是唯一一种一次抚育两个宝宝的企鹅。

洪堡企鹅（Humbolt Penguin，别名：秘鲁企鹅）生活在气候温暖的地区，比如南非。

这是一只帽带企鹅（Chinstrap Penguin，别名：纹颊企鹅）。你知道它的名字从何而来吗？它生活在南极洲的岛屿上。

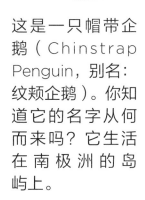

我最爱吃枪乌贼。

再见,我们明年冬天见!

生命循环，周而复始
The circle of life goes round and round

现在你知道我怎样长成一只成年企鹅了吧！

Now you know how I turned into a grown-up penguin.

结束动作：飞！

帝企鹅只有在跃出水面时才会"飞"，尽管这不是真正意义上的"飞"。它们如同一枚枚小火箭从水里发射，然后用肚皮在冰面（ice）上完美着陆。

哇！这些鲭鱼真是一顿美味的大餐。

帝企鹅是潜水高手，可以一口气在海里潜泳20分钟。

我的第一堂捕鱼课
（My first fishing lesson）

我开始学习（learn）如何潜水和捕食美味的（tasty）小鱼了。通过观察（watching）成年企鹅，我学会了捕鱼。

企鹅紧追着鱼群（shoals of fish），一抓到鱼，就一口吞进（swallow）肚子里。

看!这些企鹅正排着队(in line),等着潜入(dive into)冰冷的(icy cold)海里。

吃大餐了

企鹅的食物种类丰富多样,比如鱼类和贝类(shellfish),但它们最爱吃的就是枪乌贼(squid)和一种叫作磷虾(krill)的小生物了。

磷虾

枪乌贼

我第一次游泳
（My first swim）

换上新"泳衣"，我迫不及待地想要下海觅食了。

鳍状肢是企鹅的游泳好帮手，能帮助它们在水下快速游动。

防水羽毛

企鹅宝宝需要几周才能褪去(lose)绒毛,长出防水的羽毛,这个过程(process)就叫作换羽(fledging)。

焕然一新

成年企鹅每年也会褪去旧羽毛(feather),长出新羽毛。

长出新羽毛
（My adult feathers grow in）

5个月后，我开始长出成年企鹅才有的防水羽毛。这意味着（mean）我终于可以游泳并外出觅食了。

嘎！嘎！我饿了，快喂我好吃的！

这只企鹅已经长得和它的爸爸一样高大（big）了，但是它还无法独立捕食。

当企鹅爸爸和企鹅妈妈外出觅食时,小家伙们就挤在一起(huddle together),由几只成年企鹅照顾。和人类的幼儿园一样,企鹅也有"企鹅幼儿园"(crèche)。

企鹅宝宝小知识

- 在5个月的生长期里,企鹅宝宝大约只进食16次。

- 企鹅宝宝挤在一起,可以抱团取暖。

爸爸和妈妈的照料
(My dad and mum both look after me)

我破壳而出后,爸爸和妈妈轮流照顾我,给我喂食(feeding),还给我保暖(warm)。在它们的精心照料下,我逐渐成长,越来越健壮。

宝宝叫一叫,爸妈就找到。

吐出食物

企鹅爸爸和企鹅妈妈将食物储存在胃(stomach)里,回家后再吐出来(spit up)喂给企鹅宝宝吃。

绒毛保养

企鹅宝宝的绒毛(feather)柔软又暖和,但这种绒毛既不防水也容易脏。企鹅妈妈必须常常帮小家伙梳理绒毛,保持绒毛的洁净(clean)与蓬松(fluffy)。

我已经出生两天啦!

企鹅妈妈用大大的脚趾(toe)托起企鹅宝宝,将它与冰冷的地面(ground)隔开,给它保暖。

我要破壳而出啦
(It's time for me to hatch)

整整两个月,爸爸一直给我保暖,我将在爸爸的育儿袋中出生(born)。直到妈妈回来(return),爸爸把我交给妈妈后自己才去觅食。

咔嚓,咔嚓,咔嚓……

……我来啦!

终于自由了!

破壳而出

企鹅宝宝的喙(beak)的尖端有一个坚硬的凸起,叫作卵齿(egg tooth)。企鹅宝宝用它啄破蛋壳,然后慢慢地(slowly)爬出来。整个过程大约需要3天。

屹立风雪中

企鹅爸爸必须缓慢地（slowly）、小心翼翼地（carefully）挪动脚步，以防蛋从它的脚背上掉落（drop）。

大海太远了，企鹅妈妈还得赶好几天路才能到达。

用脚推动的旅程

企鹅妈妈肚皮（tummy）贴地，蹬脚滑行（sliding），奔向大海，这样可比走着去快多了！冬天，企鹅妈妈在海里敞开肚皮尽情吃；春天，企鹅妈妈会在企鹅蛋孵化（hatch）前赶回陆地。

爸爸，轮到你了
（Dad is left in charge）

妈妈产蛋消耗了大量体能，必须马上返回大海觅食，补充能量。妈妈只能把孵蛋的重任交给爸爸。爸爸会把蛋藏进自己的育儿袋（pouch）里并保护它，让它保持温暖（warm）和安全（safe）。

企鹅妈妈离开的这段时间，无法捕食的爸爸只能忍饥挨饿。

毛茸茸的育儿袋

请仔细观察这张图片。你能看到在企鹅爸爸松软的羽毛下面，有一颗蛋（egg）正舒服地卧着吗？

这是我的爸爸。

唔唔…… 这是我的妈妈。

我的爸爸和妈妈
（My dad and mum）

一年中的大部分时间,我的爸爸和妈妈各自生活在海（sea）里。只有到每年4月（April）的繁殖季时,它们才会回到陆地上,并通过彼此（each other）的叫声找到对方。

企鹅小知识

- 企鹅每年都会回归到同一个企鹅族群。
- 全世界大约只有40个帝企鹅族群,其中成年帝企鹅约有400,000只。
- 成年帝企鹅的身高接近3~4岁儿童的身高。

响亮的叫声

帝企鹅的叫声很响亮（loud）。这一点很重要,因为它们是通过声音而不是视觉来辨认同伴的。

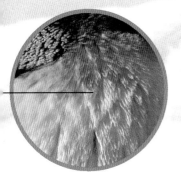

帝企鹅的耳朵（ear）藏在这片橘黄色的皮毛（fur）下面。

群居生活

在陆地上，帝企鹅（emperor penguin）喜欢聚居生活，称作族群（colony）。

翻开下一页，看看我是怎样长大的吧！

我是一只企鹅
（I'm a penguin）

我生活在寒冷的南极（Antarctic）。别看我是一只不会飞（fly）的海鸟，我游泳（swim）可棒着呢！我能从海洋（ocean）里找到美味的食物。

为了适应在海里捕食，企鹅的翅膀逐渐演化成鳍状肢（flipper），帮助它们在水下快速游动。"水中小快艇"可不是浪得虚名。

凭借长而尖锐的爪子（claw），企鹅在冰雪上也能行走自如。

18~19
我的第一堂捕鱼课

20~21
生命循环,周而复始

22~23
我世界各地的朋友

24
词汇表

和我们一起潜水,看着我们成长吧!

目录 Contents

Original Title: Penguin
Copyright © Dorling Kindersley Limited, 2004
A Penguin Random House Company

本书简体中文版授权由人民邮电出版社独家出版，仅限于中国境内（不包括香港、澳门、台湾地区）销售。未经出版者书面许可，不得以任何方式复制或发行本书中的任何部分。

4~5
我是一只企鹅

6~7
我的爸爸和妈妈

8~9
爸爸，轮到你了

10~11
我要破壳而出啦

12~13
爸爸和妈妈的照料

14~15
长出新羽毛

16~17
我第一次游泳

For the curious
www.dk.com

DK 动物成长奥秘
看！我在长大（中英双语版）

企鹅

英国 DK 公司○编

方巾予 滕懿文 岳雨燊 牛潇亦○译

鹰之舞 沈成○审

人民邮电出版社

北京

WATCH ME GROW
PUPPY

Contents

4~5
I am a dog

6~7
My dad and mum

8~9
We are three days old

10~11
Now we can see

12~13
It's time to explore

14~15
I'm three months old

16~17
Playing with dad

18~19
I'm one year old

20~21
The circle of life

22~23
My friends from around the world

24
Glossary

I am a dog

I have a wet, black nose and a long pink tongue. I love to play and chase after things. My long tail swishes back and forth when I am happy or excited.

The fur has a thick undercoat and a fine outer coat.

Labrador retriever

Retriever dogs are trained to grab things in their mouths and bring them back. They also love to run and chase after things.

Snore Snore... lots of puppies, all snuggled up to sleep.

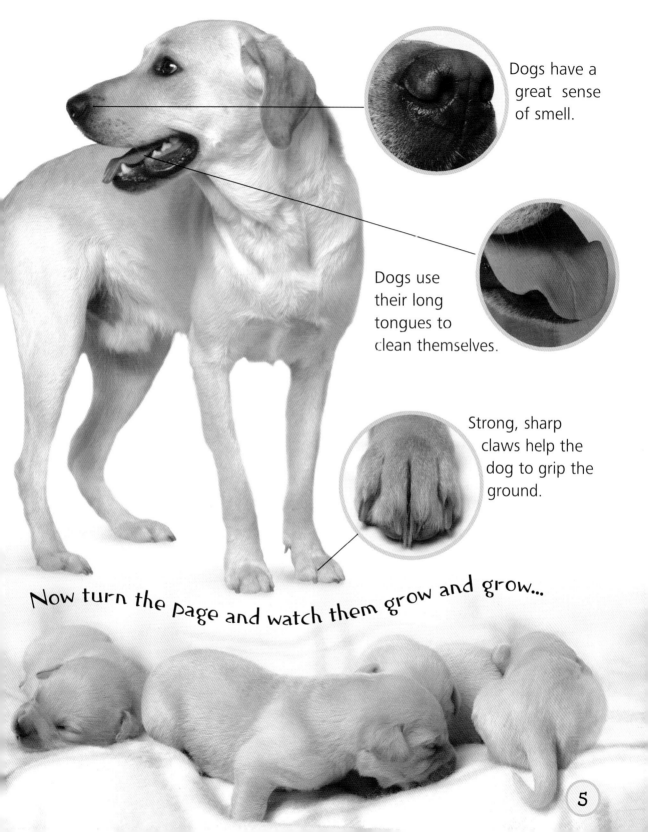

Dogs have a great sense of smell.

Dogs use their long tongues to clean themselves.

Strong, sharp claws help the dog to grip the ground.

Now turn the page and watch them grow and grow...

My dad and mum

My dad and mum live on a farm. They like to run and play together. But after my brothers, sisters and I are born, mum will look after us by herself.

This is my mum.

Mum's tummy
Can you see how big the mother dog's tummy is? The puppies have been growing inside her for eight weeks. Soon they will be born.

This is my dad.

Dog facts

- Dogs sweat through the cushioned pads of their feet.
- A dog's sense of smell is so good that a dog can smell each ingredient in a pot of stew.
- Dogs can see well in the dark. They can see some colours, but probably not red and green.
- Dogs bark to tell other dogs that something is happening.

After a bath, dad likes to shake himself dry!

We are three days old

Mum gave birth to us in a warm, safe place. We can't see or hear, but we can smell and wriggle around. We spend most of our time eating, sleeping, and keeping warm and cosy.

The puppies nurse by gently pushing mum's breasts with their paws.

It's a tight squeeze

The puppies squirm and shove to find a teat and begin drinking their mother's milk. They will nurse every three or four hours.

No pushing, there's room for everyone.

Now we can see

Our eyes opened when we were nine days old. Now we are learning to walk. Mum wants us to stay close, but we're curious about the world and want to begin exploring.

The puppies still spend most of their time sleeping.

Learning to walk

The puppies are still unsteady on their feet, but they will soon be strong enough to run and play.

Growing up is hard work!

It's time to explore

I'm six weeks old and big enough to explore the house on my own. There are so many interesting things to see! It's great fun finding new toys to play with.

Learning by playing

The puppies have sharp teeth now and they chew and play with everything.

Puppies say "hello" to new friends by sniffing them.

Sniff sniff... come on teddy, let's play!

Dinner time

The puppies can eat solid food. They are still nursing, but they nurse less each day. Soon they will be weaned and eat only solid food.

I'm three months old

It's time for me to start exploring the forest. With my long snout I can smell all the insects and plants in the woods.

This puppy has found an exciting smell, buried deep in the leaves.

On the scent

Dogs have a much better sense of smell than people. Dogs use smell to find food and as a way to recognise other dogs and humans.

Playing with dad

I am six months old and now I am almost as big as dad. When I play tug-of-war with dad, he growls to let me know that he is still in charge of our family.

Come on dad! Let me have the stick!

Dogs of all ages learn and grow by playing games.

Super swimmers

Most dogs are very good swimmers, but some like the water better than others. All dogs use the same swimming stroke – the dog paddle.

Grrrrr...

I'm one year old

While I've been busy growing up, our family has grown bigger. Mum and dad have had six new puppies. These are my new brothers and sisters.

Many dogs enjoy living in a big, friendly family group like this one.

My mum

Busy mum
Adult female dogs can have two litters of puppies each year, starting when they are just six months old.

The circle of life goes round and round

Now you know how I turned into a grown-up dog.

My friends from around the world

My doggy friends from around the world have many jobs to do, so they come in lots of sizes and shapes.

The Kerry Blue Terrier comes from Ireland and loves the water.

Old English Sheepdogs love to help farmers to look after their sheep.

Bassett Hounds come from France, where they were once used as hunting dogs.

The Chihuahua comes from Mexico and is the world's smallest dog.

Afghan Hounds have been used as hunting dogs for thousands of years.

Woof woof – I help move sheep around the farm.

This Pomeranian is 30 cm (1 ft) tall.

Doggy facts

- Dogs can hear many sounds that humans cannot, such as insects flying.
- The first animal in space was a Russian dog named Laika, sent into orbit in 1957.

Glossary

Retrieve
To find something and to bring it back. Labradors are good retrievers.

Growl
A warning noise that tells other dogs or people to watch out.

Nurse
When a baby animal, such as a puppy, drinks its mother's milk.

Snout
The name for the nose and mouth of some animals, including dogs.

Fur
The soft hair on a dog that keeps it warm and protects it.

Wean
When a baby stops drinking milk and starts to eat solid food.

Acknowledgements
感谢以下人员及机构提供图片：
Key: t = top, b = bottom, l = left, r = right, bkgrd = background, c = centre

4 Alamy Images: Stefanie Krause-Wieczorek cl. 7 Corbis: Tom Stewart br. 19 Science Photo Library: Renee Lynn tr. 24 Ridgeway Labradors: br.

其他图片版权属于多林·金德斯利公司。

欲了解更多信息请访问DK Images网站。在此特别感谢Helen和Stephen Harvey，谢谢他们给予我们充足的时间和耐心，并且将狗狗和房子借给我们用于拍摄；同时也感谢来自英国牛津郡的Ridgeway Labradors，以及我们的狗狗朋友们：Chloe、Phoebe、Timmy、TJ、Harvey和Roxy，多亏了它们的帮助与配合，这本书才得以完成。

词汇表 Glossary

寻回
Retrieve

找到猎物并叼回来的行为，拉布拉多猎犬是优秀的寻回犬。

低吼
Growl

动物通过低声咆哮来发出警告。

哺乳
Nurse

动物（比如狗妈妈）产下小宝宝后，用自己的乳汁哺育它们的行为。

口鼻部
Snout

某些动物的鼻子和嘴巴的统称。

皮毛
Fur

动物体表一层柔软的毛发，用来御寒保暖。

断奶
Wean

婴儿或幼小的哺乳动物不再吃母乳，改吃固体食物的阶段。

致谢 Acknowledgements

感谢以下人员及机构提供图片：

Key: t = top, b = bottom, l = left, r = right, bkgrd = background, c = centre

4 Alamy Images: Stefanie Krause-Wieczorek cl. 7 Corbis: Tom Stewart br. 19 Science Photo Library: Renee Lynn tr. 24 Ridgeway Labradors: br.

其他图片版权属于多林·金德斯利公司。

欲了解更多信息请访问DK Images网站。在此特别感谢Helen和Stephen Harvey，谢谢他们给予我们充足的时间和耐心，并且将狗狗和房子借给我们用于拍摄；同时也感谢来自英国牛津郡的Ridgeway Labradors，以及我们的狗狗朋友们：Chloe、Phoebe、Timmy、TJ、Harvey和Roxy，多亏了它们的帮助与配合，这本书才得以完成。

吉娃娃（Chihuahua），是世界上体形最小的狗。

阿富汗猎犬（Afghan Hound）是一种古老的犬种，几千年前就跟随猎人狩猎。

汪汪！我负责看管农场的羊群。

这只博美的身高只有30厘米。

狗的小知识

🐕 狗的听力十分灵敏，能听到很多人类听不到的声音，比如昆虫振翅时发出的声音。

🐕 1957年，苏联将一只名叫莱卡的小狗送入了太空，它成为了第一个乘坐卫星环绕地球的动物。

我世界各地的朋友
My friends from around the world

我的朋友来自世界各地（around the world），它们体形不同、长相各异，而且它们承担着不同的职责。

凯利蓝㹴（Kerry Blue Terrier）又称爱尔兰㹴，它的家乡在爱尔兰的凯里郡。它特别喜欢玩水（water），善于寻回掉落在水中的东西。

英国古典牧羊犬（Old English Sheepdog）是英国最古老的牧羊犬犬种之一，顾名思义，它们能帮助人类看管羊群（sheep）。

巴吉度猎犬（Bassett Hound）原产法国（France），它可是追捕（hunting）猎物的高手！

生命循环，周而复始
The circle of life goes round and round

现在你知道我怎样长成一只大狗了吧！

Now you know how I turned into a grown-up dog.

狗都喜欢生活在这样友好（friendly）的大家庭里。

这是我的妈妈

忙碌的妈妈

母狗（female dog）6个月大时就发育成熟，可以当妈妈了。每年它们可以生育两次。

我一岁了
（I'm one year old）

随着我的成长（grow up），我的家族也日渐壮大。爸爸和妈妈又有了6个宝宝。瞧，它们都是我的家人！

我的表哥　　我的姑姑　　这是我　　我的爸爸

游泳健将

大多数狗是游泳健将（good swimmer）。它们使用同一种泳姿——狗刨式（the dog paddle）。

呜嗷……

和爸爸玩耍
（Playing with dad）

6个月大的我，体形已经和爸爸差不多了。但当我们拔河（tug-of-war）时，爸爸会发出低吼，表示它仍是我们家的"老大"！

快点！爸爸，把棍子给我！

不管在哪个年龄段（ages），狗总是在玩耍（playing games）中成长起来的。

关于嗅觉

我有着比人类更灵敏的嗅觉。它除了能帮我找寻食物,还能帮我区分(recognise)人类和同伴。

我3个月大喽
（I'm three months old）

这时候的我，已经可以去森林（forest）里探险了。我的小鼻子，能够帮我分辨出森林里的昆虫（insect）和植物（plant）的气味。

这只小狗嗅到了树叶（leaf）深处的某种气味，这令它非常兴奋。

吃大餐啦！

狗宝宝现在已经可以尝尝固体食物（solid food）了，但仍然需要喝奶，只是每天喝奶的次数正在减少。很快它们就可以彻底断奶（wean），改吃固体食物了。

泰迪熊，让我们一起玩吧！

是时候去探索世界了
（It's time to explore）

出生6周后，我终于可以独自探索世界了。有那么多有趣的事情（interesting thing）等着我去发现，我真是太兴奋啦！如果能找到新玩具（toy）一起玩耍，那我就会高兴一整天呢！

寓教于乐

狗宝宝已经长出了尖锐的小牙齿（teeth），不仅可以咀嚼（chew）食物，还可以叼着东西玩耍！

狗宝宝通常用嗅嗅（sniffing）对方的方式来和朋友打招呼。

嗅嗅这边，嗅嗅那边，快来！

学会行走

虽然这时候的狗宝宝还站不稳（unsteady），但它们很快就会强壮起来，可以尽情地奔跑和玩耍。

成长真不容易！

现在我们能看见啦
（Now we can see）

出生9天以后，我们第一次睁开双眼看见这个精彩的世界。接着我们开始学习走路（walk），虽然妈妈希望我们寸步不离（stay close）地跟着，但好奇的（curious）我们已经迫不及待地想要开启探索之旅啦！

对于这个阶段的狗宝宝来说，一天中的大部分时间都在睡觉（sleeping）。

太挤了!

狗宝宝扭动着(squirm)、争先恐后地寻找妈妈的乳头,然后贪婪地吮吸乳汁。它们每隔3~4小时(every three or four hours)就要喝一次奶。

别挤啦,大家都有份。

我们出生3天了
（We are three days old）

我们出生在一个温暖（warm）而安全（safe）的地方。刚出生时，我们既看不见也听不见，但我们可以闻到气味，还能缓慢地四处蠕动（wriggle around）。我们每天要做的事就是惬意地吃饱睡足。

在喝奶时，小狗会用爪子（paw）轻轻地（gently）按压着妈妈的乳房。

这是我的爸爸。

洗完澡后，爸爸喜欢甩干身上的水。

狗的小知识

- 狗通过舌头和脚底肉垫上的汗腺来分泌汗液。
- 狗的嗅觉非常灵敏，甚至能够辨别出一锅肉汤里的每种配料。
- 狗有敏锐的暗视力。当天色比较暗的时候，狗的视力反而比人好很多。狗能看到除了红和绿外的很多颜色。
- 狗通过吠叫来提醒或警示同伴。

我的爸爸和妈妈
（My dad and mum）

我的爸爸和妈妈生活在农场（farm）。它们喜欢在一起奔跑、嬉戏。我和兄弟姐妹出生后，妈妈就会独自肩负起照顾（look after）我们的重任。

这是我的妈妈。

妈妈的肚子

看到我们的妈妈的肚子（tummy）有多大（big）了吧！我们已经在妈妈肚子里成长8周了！很快，我们就会出生啦！

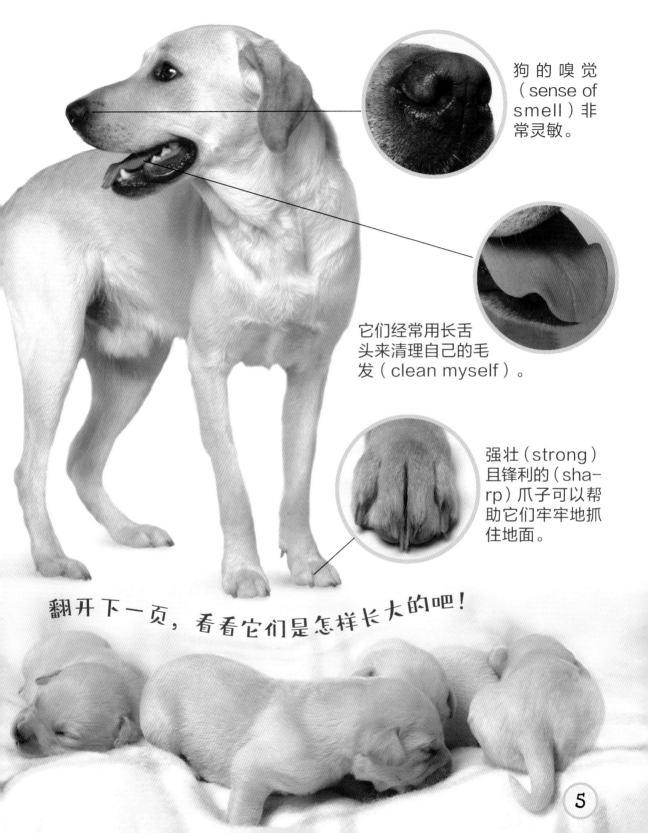

狗的嗅觉（sense of smell）非常灵敏。

它们经常用长舌头来清理自己的毛发（clean myself）。

强壮（strong）且锋利的（sharp）爪子可以帮助它们牢牢地抓住地面。

翻开下一页，看看它们是怎样长大的吧！

我是一只狗
（I am a dog）

黑鼻子（black nose），湿漉漉；长舌头（long tongue），粉（pink）嘟嘟；爱玩耍，爱追逐，这就是我——一只狗。每当我感到开心或兴奋时，我就会摇摆我的尾巴。

我厚厚的（thick）皮毛（fur），像一件柔软的外套。

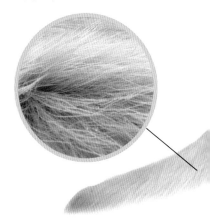

拉布拉多猎犬

拉布拉多猎犬可以用嘴巴（mouth）叼住（grab）抛出的物品并带回来。所以，它们也被称为寻回犬。它们喜欢奔跑（run）和追逐（chase）。

呼噜呼噜……我们紧紧依偎着进入梦乡。

12~13
是时候去探索世界了

14~15
我3个月大喽

16~17
和爸爸玩耍

18~19
我一岁了

20~21
生命循环，周而复始

22~23
我世界各地的朋友

24
词汇表

Original Title: Puppy
Copyright © Dorling Kindersley Limited, 2005
A Penguin Random House Company

本书简体中文版授权由人民邮电出版社独家出版,仅限于中国境内(不包括香港、澳门、台湾地区)销售。未经出版者书面许可,不得以任何方式复制或发行本书中的任何部分。

For the curious
www.dk.com

目录 Contents

4~5
我是一只狗

6~7
我的爸爸和妈妈

8~9
我们出生3天了

10~11
现在我们能看见啦

DK 动物成长奥秘
看！我在长大（中英双语版）

小·狗

英国 DK 公司◎编
王景乐宁 梁逸凡 Pei Qi Alexander Liu 邱梓涵◎译
鹰之舞 沈成◎审

人民邮电出版社
北京

DK WATCH ME GROW
RABBIT

Contents

4~5
I'm a rabbit

6~7
My dad and mum

8~9
Here is my home

10~11
Inside my family's nest

12~13
I'm getting curious

14~15
I'm four weeks old

16~17
I can find my own dinner

18~19
Now I'm grown up

20~21
The circle of life

22~23
My friends from around the world

24
Glossary

I'm a rabbit

I have two large ears and a fluffy tail. My strong back legs help me to leap and run, and I am covered in soft fur to keep me warm.

A great sense of smell helps rabbits to sniff out danger.

Strong claws are good for digging.

This is my family.

Rabbits have many layers of thick fur.

Running away
Rabbits can run about as fast as you can ride your bicycle. This helps them outrun their enemies.

Wake up everyone, it's time to get hopping.

My dad and mum

My dad and mum made their home under a large tree. They live with other rabbit families in a series of underground tunnels called a warren.

This is dad.

This is mum.

Saying hello

When rabbits meet, they say hello to each other by sniffing and touching noses.

This hole is an entrance tunnel.

Family facts

- Rabbit mums and dads do not stay together after the babies are born.
- Rabbits work as a team to build their tunnels.
- A warren in England was found with 1,027 entrance and exit holes. Over 400 rabbits lived in it.

Here is my home

After they meet, mum and dad dig a big, cosy chamber where we will be born. This is our nest.

Shopping trip
Up on the surface, rabbits collect grass, feathers, and fur to make a soft nest.

Can you see the cosy bed mum has made?

I'm on the lookout.

The warren is a safe place to live. Larger animals find it hard to squeeze through the tunnels.

Inside my family's nest

When I am born, my eyes are shut tight and I don't have any fur. We stay warm by huddling close together in the cosy nest mum has built.

Day one... day two...

day three.

Now I'm two weeks old.

All grow bigger!

At first the bunnies are very weak. They can hardly move at all. But soon they can move around. After 10 days they open their eyes.

Lullaby bunny

When they are not eating, the babies sleep for almost all of the time. Growing fast is very tiring.

After two weeks, the bunnies have thick fur and they spend more time awake.

I'm getting curious

After about three weeks, we like to run and jump around inside the warren. We can't wait to start exploring the big world outside.

Mother's milk
Mother rabbits feed their baby bunnies with milk. This is called nursing. The bunnies are fed only once each day.

Rabbit facts

🐇 Rabbits spend most of their day underground. They come out to eat at sunset and at dawn.

🐇 Rabbits can hear many sounds that humans cannot.

🐇 The deepest warren tunnels ever found were dug 9 metres (27 ft) underground.

Now the bunnies are full and sleepy.

I'm four weeks old

It's time for me to hop out of my burrow. My brothers and sisters and I stay close to the entrance, just in case we need to **ZOooom** back inside!

Rabbits' ears stick up high to listen for danger.

The warm sand is a comfy place for the rabbits to sit.

Take care!

Outside the safety of the warren, the bunnies are always on the lookout for danger. Bunnies are a meal for animals like eagles, foxes, and weasels.

Eagle

Fox

I'm so shy, I think I'll hide!

It's back down the tunnel when we get a fright!

The bunny's eyes quickly get used to bright daylight.

I can find my own dinner

Now I am old enough to look for food on my own. Green, leafy plants are my favourite food. I find yummy plants to eat growing in fields and in gardens.

This looks like a good place to eat dinner.

Favourite food

Rabbits eat many different kinds of plants. Most rabbits love to eat dandelion, carrot leaves, hay, and parsley.

Dandelion

Rabbits don't eat snails!

Rabbits will eat leafy plants like this Swiss chard wherever they find them — even in your garden!

Now I'm grown up

After just eight weeks I am all grown up. I spend my time running in the fields, looking for good things to eat. Soon it will be time to start a family of my own.

Rabbits smell plants to see if they are good to eat.

Bath time
Rabbits spend a lot of time washing their fur. This is called grooming.

The rabbits' noses tell them that these buttercups are not good to eat.

Family life
Adult rabbits live together in a group. When new rabbits are born, they stay in the group.

The circle of life goes round and round

Now you know how I turned into a grown-up rabbit.

My friends from around the world

Can you see why this rabbit is called a Blue Dwarf?

The Jackrabbit lives in the desert. Its big ears help it to stay cool in the desert sun.

This Angora Rabbit has a lot of fur, and even funny tufts of fur on the tips of its ears!

The Mountain Hare has huge feet, which help it to run in the snow.

My fluffy, furry rabbit friends come in all sizes and colours.

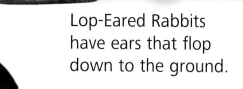

Dutch Rabbits are one of the most popular pet rabbits. They are known for being very gentle.

Lop-Eared Rabbits have ears that flop down to the ground.

The tiny Pika is the size of a hamster!

The Lion-Head Rabbit has a mane of fur on top of its head.

The Mountain Hare is white so its enemies can't see it in the snow.

Rabbit facts

- A rabbit's teeth grow throughout its entire life. Rabbits need to chew wood to keep their teeth short.
- When rabbits are happy they like to jump around and leap about.
- Rabbits are the most active at dawn and at twilight.

Glossary

Warren
A system of underground tunnels that rabbits live in.

Claw
A sharp nail on the rabbit's toe that helps the rabbit to dig.

Nest
A place inside the warren where rabbit babies are born.

Bunny
The name for a baby rabbit that is less than one month old.

Litter
A group of baby rabbits all born to the same mother.

Grooming
When the rabbit washes its fur to keep it clean and free of bugs.

Acknowledgements

感谢以下人员及机构提供图片:

(Key: a=above; c=centre; b=below; l=left; r=right; t=top)
1 Warren Photographic: Jane Burton c, bl. 2-3 Bruce Coleman Ltd: William S. Paton bkg&c. 2 DK Images: Jane Burton b. 3 Oxford Scientific Films: Jorge Sierra Antinolo &c. 4-5 DK Images: Geoff Dann. 4 DK Images: Dave King l; Steven Moore Photography r; Bruce Coleman Ltd: William S. Paton b.
5 Bruce Coleman Ltd: William S. Paton br; FLPA - Images of nature: David Hosking tr; Steven Moore Photography tl; Warren Photographic: Jane Burton bl. 6-7 Corbis: Terry W. Eggers t; DK Images: Steve Shott b; 6 DK Images: Steve Shott c; Jane Burton cr; Bruce Coleman Ltd: Colin Varndell bl. 7 Corbis: Phillip Gould tc. 8-9 Corbis: Terry W. Eggers t; DK Images: Steve Shott b. 8 Corbis: Phillip Gould; Oxford Scientific Films cl. 9 DK Images: Jane Burton tl. 10-11 Oxford Scientific Films.10 Bruce Coleman Ltd: Jane Burton tl, c; Warren Photographic: Jane Burton tr. 11 Bruce Coleman Ltd: Jane Burton tr.
12 DK Images: Steve Shott l. 12-13 Oxford Scientific Films: Maurice Tibbles.
14-15 Bruce Coleman Ltd: William S. Paton. 15 DK Images: Frank Greenaway rca; Jerry Young rc; Oxford Scientific Films: br. 16-17 Steven Moore Photography. 16 DK Images: Ian O'Leary l; FLPA - Images of nature: Tony Hamblin br. 17 Royalty Free Images: Photofrenetic/ Alamy tr; DK Images: Jacqui Hurst br. 18-19 Bruce Coleman Ltd: Colin Varndell. 18 FLPA - Images of nature: Tony Hamblin br. 19 Oxford Scientific Films: Mike Powles tr. 20 Bruce Coleman Ltd: Jane Burton tc, trb; Corbis: Tony Hamblin bl; DK Images: Barrie Watts bc; Geoff Dann tl; Jane Burton cl; Warren Photographic: Jane Burton tr, cr, crb, c. 21 Steven Moore Photography bkg; Oxford Scientific Films: Jorge Sierra Antinolo c. 22 DK Images: Jane Burton tl; Steve Shott cl; Oxford Scientific Films: Paul Berquist tr; Warren Photographic: Jane Burton c. 22-23 Powerstock: Superstock. 23 Corbis: George D. Lepp trb; DK Images: Steve Shott tr; FLPA - Images of nature: Terry Whittaker cla; Steven Moore Photography: br; Warren Photographic: Jane Burton cl. 24 Bruce Coleman Ltd: Jane Burton cr, bl; DK Images: Steve Shott tl, cl; Steven Moore Photography: tr.

其他图片版权属于多林·金德斯利公司。
欲了解更多信息请访问DK Images网站。

词汇表 Glossary

兔子洞
Warren
兔子居住的地下巢穴。

爪子
Claw
兔子的脚趾上有尖尖的趾甲，可以帮助它挖洞。

小窝
Nest
兔子洞里属于每个兔子家庭的小窝，是兔宝宝出生的地方。

兔宝宝
Bunny
出生不到一个月的兔子。

一窝幼崽
Litter
由同一个兔妈妈生出来的一窝兔宝宝。

梳毛
Grooming
兔子舔洗自己的皮毛，以除虫和保持清洁。

致谢 Acknowledgements

感谢以下人员及机构提供图片：

(Key: a=above; c=centre; b=below; l=left; r=right; t=top)

1 Warren Photographic: Jane Burton c, bl. 2-3 Bruce Coleman Ltd: William S. Paton bkg&c. 2 DK Images: Jane Burton b. 3 Oxford Scientific Films: Jorge Sierra Antinolo br. 4-5 DK Images: Geoff Dann. 4 DK Images: Dave King l; Steven Moore Photography r; Bruce Coleman Ltd: William S. Paton b. 5 Bruce Coleman Ltd: William S. Paton br; FLPA - Images of nature: David Hosking tr; Steven Moore Photography tl; Warren Photographic: Jane Burton bl. 6-7 Corbis: Terry W. Eggers t; DK Images: Steve Shott b; 6 DK Images: Steve Shott c; Jane Burton cr; Bruce Coleman Ltd: Colin Varndell bl. 7 Corbis: Phillip Gould tc. 8-9 Corbis: Terry W. Eggers t; DK Images: Steve Shott b. 8 Corbis: Phillip Gould; Oxford Scientific Films cl. 9 DK Images: Steve Shott tlb. 10-11 Oxford Scientific Films.10 Bruce Coleman Ltd: Jane Burton tl, c; Warren Photographic: Jane Burton tr. 11 Bruce Coleman Ltd: Jane Burton tr. 12 DK Images: Steve Shott l. 12-13 Oxford Scientific Films: Maurice Tibbles. 14-15 Bruce Coleman Ltd: William S. Paton. 15 DK Images: Frank Greenaway rca; Jerry Young rc; Oxford Scientific Films: br. 16-17 Steven Moore Photography. 16 DK Images: Ian O'Leary l; FLPA - Images of nature: Tony Hamblin br. 17 Royalty Free Images: Photofrenetic/ Alamy tr; DK Images: Jacqui Hurst br. 18-19 Bruce Coleman Ltd: Colin Varndell. 18 FLPA - Images of nature: Tony Hamblin br. 19 Oxford Scientific Films: Mike Powles tr. 20 Bruce Coleman Ltd: Jane Burton tc, trb; Corbis: Tony Hamblin bl; DK Images: Barrie Watts bc; Geoff Dann tl; Jane Burton cl; Warren Photographic: Jane Burton tr, cr, crb, c. 21 Steven Moore Photography bkg; Oxford Scientific Films: Jorge Sierra Antinolo c. 22 DK Images: Jane Burton tl; Steve Shott cl; Oxford Scientific Films: Paul Berquist tr; Warren Photographic: Jane Burton c. 22-23 Powerstock: Superstock. 23 Corbis: George D. Lepp trb; DK Images: Steve Shott tr; FLPA - Images of nature: Terry Whittaker cla; Steven Moore Photography: br; Warren Photographic: Jane Burton cl. 24 Bruce Coleman Ltd: Jane Burton cr, bl; DK Images: Steve Shott tl, cl; Steven Moore Photography: tr.

其他图片版权属于多林·金德斯利公司。欲了解更多信息请访问DK Images网站。

我毛茸茸的兔子朋友的体形和毛色各不相同。

荷兰兔（Dutch Rabbit）是最受欢迎的宠物兔之一，以温顺著称。

垂耳兔（Lop-Eared Rabbit）的耳朵长到会拖在地上。

鼠兔的个头和仓鼠一般大！

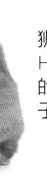

狮子兔（Lion-Head Rabbit）的头顶长着狮子似的鬃毛。

雪兔（Mountain Hare）是白色的，天敌很难在雪地里发现它们。

兔子小知识

🐰 兔子的牙齿会持续不断地生长，所以兔子需要啃木头来磨短牙齿。

🐰 兔子开心时喜欢上蹿下跳。

🐰 兔子在黎明和黄昏时最活跃。

我世界各地的朋友
My friends from around the world

你知道为什么这只兔子被叫作"蓝侏儒"（Blue Dwarf）吗？

杰克兔（Jackrabbit）生活在沙漠（desert）里。大大的耳朵能帮它在沙漠的烈日下保持凉爽。

这只安哥拉兔（Angora Rabbit）的毛发又多又长，耳尖上还有一撮滑稽的软毛。

雪兔长着大大的脚掌，便于它在雪地里跑跳。

再见,我要跳走喽。

生命循环，周而复始
The circle of life goes round and round

现在你知道我怎样长成一只成年兔子了吧！

Now you know how I turned into a grown-up rabbit.

我的鼻子提醒我,这些毛茛(buttercup)是不可以吃的。

家庭生活

成年兔子是群居生活的。兔宝宝出生后也生活在这个群体里。

我终于长大了
（Now I'm grown up）

仅仅8周的时间，我就可以长成一只成年兔子。我成天在田野里跑来跑去，寻找美味的食物。很快，我也要组建自己的家庭（family）了。

兔子会通过嗅觉，来判断植物是否可以食用。

梳毛时间

我会花很多时间来舔洗毛发，这叫作梳毛（grooming）。

最喜欢的食物

兔子会吃很多不同种类的植物。大部分兔子喜欢吃蒲公英（dandelion）、胡萝卜叶、干草（hay）和欧芹（parsley）。

蒲公英

兔子不吃蜗牛！

我爱吃多叶植物，比如莙荙菜，即使它生长在你的菜园里，我也照吃不误！

我能自己觅食啦
（I can find my own dinner）

现在我已经足够强壮了，能够独自外出觅食。绿色的多叶植物（leafy plant）是我的最爱。不管是在田野（field）里还是花园（garden）里，我都能找到可口的食物。

这里看起来是一个享用美食的好地方。

小心捕食者！

一旦离开兔子窝，兔宝宝就会随时保持警觉。它们是鹰（eagle）、狐狸（fox）和鼬（weasel）等捕食者眼中的美餐。

鹰

狐狸

好害怕呀，我还是躲起来吧！

我的眼睛能快速适应明亮的光线。

不小心一害怕，就会躲回窝里！

我4周大了
（I'm four weeks old）

我可以跳出去探索家外边的世界啦！我和我的兄弟姐妹都待在洞口（entrance）附近，一有情况我们就会"嗖"地一下躲回兔子窝里。

我的耳朵竖得高高的，倾听一切可疑的声音，以防危险突然到来。

趴在暖暖的沙子上好舒服啊！

兔子小知识

🐰 一天中的大部分时间兔子都待在地下。它们只在黎明和黄昏的时候才出来觅食。

🐰 兔子可以听见人类无法听见的很多声音。

🐰 迄今发现的最深的兔子洞深达9米。

我们吃饱了，现在好困呀！

好奇宝宝就是我
（I'm getting curious）

大约3周后，我们就能在窝里蹦来蹦去了。真是等不及去探索外面的（outside）世界了。

妈妈的乳汁

妈妈用自己的乳汁（milk）来喂养（feed）兔宝宝，这种行为叫作哺乳（nursing）。兔宝宝一天只喝一次奶。

兔宝宝睡呀睡

兔宝宝整天几乎除了吃就是睡。长这么快可是一件很辛苦（tiring）的事呢！

两周之后，兔宝宝长出了厚厚的绒毛，清醒的（awake）时间更长了。

在我家的小窝里
（Inside my family's nest）

我刚出生时，眼睛（eye）闭得紧紧的（tight），身上也没有长毛。在妈妈修建的舒服的小窝里，我和我的兄弟姐妹挤成一团取暖。

第一天

第二天

第三天

现在我已经有两周大了。

兔宝宝长呀长

起初，兔宝宝都很虚弱（weak），几乎无法移动，但是它们很快就可以动来动去了。出生大约10天后，兔宝宝才能睁开（open）眼睛。

我正在站岗放哨。

兔子洞是个很安全的（safe）地方，那些个头大的动物可没法挤进这些狭窄的洞穴哟！

这里是我家
（Here is my home）

爸爸妈妈相爱后，会在地下挖一个又大又舒适的"房间"，并在这里生下我，这是我们的小窝（nest）。

采购之旅

兔子会去洞外收集草（grass）、羽毛（feather）和动物毛发来建造一个松软的（soft）小窝。

你能找到我妈妈做的那张舒服的大床吗？

这个洞是兔子洞的一个入口。

家族小知识

- 兔宝宝出生后,兔妈妈和兔爸爸并没有在一起生活。
- 兔子会团结协作来修建它们的家——地下洞穴。
- 在英格兰,人们发现了一个有1027个出入洞口的巨大兔子洞,里面生活着超过400只兔子。

我的爸爸和妈妈
（My dad and mum）

我的爸爸和妈妈在一棵大树下安家。我们和其他兔子家庭一起生活在地下的洞穴（tunnel）里，这些洞穴被人们叫作兔子洞。

这是我爸爸。

这是我妈妈。

打个招呼吧！

我们见面时，会用鼻子（nose）闻一闻、碰一碰对方，来相互打招呼。

我身上长有厚厚的绒毛。

快逃呀!

我奔跑的速度和人类骑自行车（bicycle）的速度差不多，所以我能逃过敌人（enemy）的围追堵截。

起床啦，大家一起来蹦蹦跳吧！

我是一只兔子
（I'm a rabbit）

我有两只长耳朵（ear）和一根毛茸茸的短尾巴（tail）。我的后腿（leg）很强壮，所以我跑（run）得快、跳（jump）得高。我全身长满细密柔软的毛，让我感觉暖融融的（warm）。

敏锐的嗅觉能帮助我提前觉察危险。

强壮有力的爪子使我成为挖洞能手。

这是我的家人。

14~15
我4周大了

16~17
我能自己觅食啦

18~19
我终于长大了

20~21
生命循环,周而复始

22~23
我世界各地的朋友

24
词汇表

目录 Contents

Original Title: Rabbit
Copyright © Dorling Kindersley Limited, 2004
A Penguin Random House Company

本书简体中文版授权由人民邮电出版社独家出版，仅限于中国境内（不包括香港、澳门、台湾地区）销售。未经出版者书面许可，不得以任何方式复制或发行本书中的任何部分。

4~5
我是一只兔子

6~7
我的爸爸和妈妈

8~9
这里是我家

10~11
在我家的小窝里

12~13
好奇宝宝就是我

For the curious
www.dk.com

DK 动物成长奥秘

看！我在长大（中英双语版）

兔子

英国 DK 公司 ◎ 编
李渔 李白 王子戈 梁誉天 ◎ 译
鹰之舞 沈成 ◎ 审

人民邮电出版社
北京